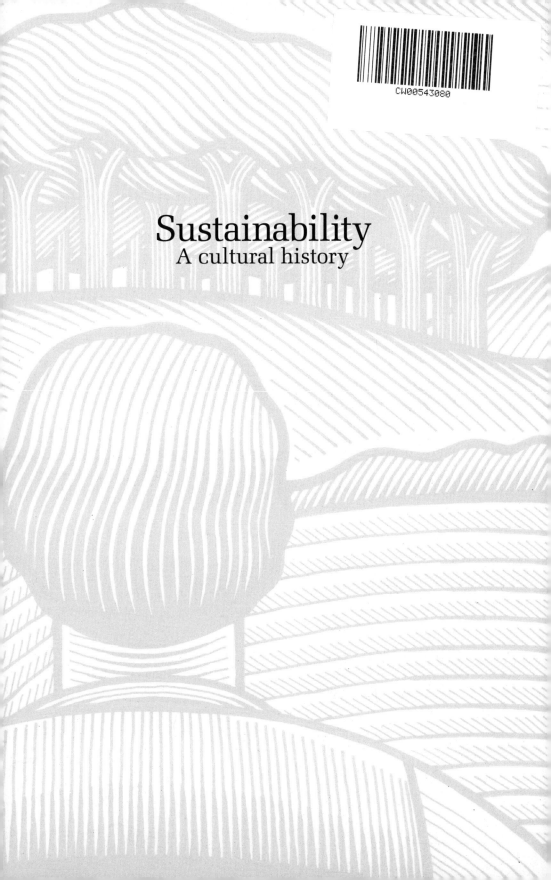

Sustainability
A cultural history

Sustainability
A cultural history

ULRICH GROBER

translated by Ray Cunningham

green books

This English edition first published in 2012
by Green Books
Dartington Space, Dartington Hall
Totnes, Devon TQ9 6EN, UK

First published in German in 2010, under the title
Die Entdeckung der Nachhaltigkeit – Kulturgeschichte eines Begriffs

Original German text © 2010 by Verlag Antje Kunstmann, Munich

English translation © Ray Cunningham

Printed on Corona Natural 100% recycled paper
by T J International Ltd, Padstow, Cornwall, UK

ISBN 978 0 85784 045 5

The translation of this book from German into English was
generously funded by Ernst Basler + Partner, Zollikon, Switzerland,
and by Hatzfeldt-Foundation, Friesenhagen, Germany.

Contents

"In an age in which we are denuding the resources of the planet as never before and endangering the very future of humanity, sustainability is the key to human survival."

Christopher Weeramantry, former Vice-President of the International Court of Justice.

Foreword

The concept of sustainability – this often overused and misunderstood term – is not an invention of the late 20th century. It is deeply rooted in many previous cultures, and in Europe can be traced back over many centuries. This is the surprising, yet entirely convincing, thesis of Ulrich Grober's book. It reveals the stark contrast between the perspectives and practices of pre-industrial eras and those of our present age, with its oft-repeated mantras of 'more is better' and 'growth is essential'.

At its core, sustainability relates to the basic human need to maintain and to nurture the conditions on which life depends. Our forbears knew this truth and acted accordingly: they understood that there can be no development without sustainability. Yet modern industrialism has become seduced by a mirror image of this truth, with the idea of linear progress based on growth in what is now clearly understood to be a finite world. This belief has become so all-pervasive that most politicians and corporate managers perceive sustainability to be the antithesis of development, a threat to progress and well-being. To them, growth is a *sine qua non* for development of any kind whatsoever.

This book explores the profound errors that are encapsulated in this belief. It demonstrates clearly that the sustainability perspective is the *only* realistic approach to social and economic development. The growth perspective, by contrast, reflects an irrational, illusory and – we could even say – insane state of mind.

The fact that to criticise the growth paradigm is viewed as extreme, idealistic or irrational is evidence that we live in a perverse world. And something is profoundly wrong when economists and politicians assume growth to be a normal and indispensable part of even 'sustainable development'.

Foreword

The German psychoanalyst Arno Gruen describes the 'insanity of normality' (1988) as a self-destructive contempt for humanity that has befallen modern society. The normal turns into the abnormal and vice versa, as we lose our sense of what is right and what is wrong. In advanced capitalism, competition is perceived as superior to cooperation, the market as more important than community, growth as implicit in development, and money as more real than people and their needs. This 'insanity of normality' is particularly visible in the way our governments are responding to the current financial crisis. Instead of controlling financial markets, they allow financial markets to control them – and us all. Markets seem to possess an overweening power and reality that compels us to conform to their needs rather than ours. Though this is madness, there is method in it: the victims, not the culprits, pay the price for greed, corruption and incompetence. It is clear that the issue of sustainability has never been more topical than it is today.

Grober quotes Joachim Heinrich Campe (the teacher of the great Alexander von Humboldt), who, in 1809, in the *Encyclopedia of the German Language*, defines the core of sustainability as "that which one holds on to when nothing else holds any longer". And he refers to the Club of Rome report *The Limits to Growth*, which states: "We are searching for a model output that represents a world system that is: 1) sustainable without sudden and uncontrollable collapse; and 2) capable of satisfying the basic material requirements of all of its people."

Two possible outcomes are presented in this book: sustainability or collapse. If it is correct, as many say, that we are now experiencing the large-scale collapse of financial, economic and governmental systems – in addition to ecological collapse – then the only option that remains for us is to find a way to achieve sustainability. It can anchor us in much the same way as freedom, equality and justice have anchored and shaped civil society. In fact, sustainability is now a prerequisite for freedom, equality and justice – their true common anchor.

It is for this reason that the much-discussed 'three pillars' model of sustainability – which suggests that ecological, economic and social systems are of equal importance – is so flawed. The ecological system is fundamental, as this book so impressively demonstrates. Ecology must always take precedence over economy, not the other way round. Economy carries the nomos within it, i.e. the measuring of the house, while ecology invokes the logos, i.e. the idea, vision and spirit of the house. The logos must instruct the nomos as to how the house should be built, and what it should look like.

History has shown that the vision of a sustainable society is not an empty dream, for such societies have existed in the past: it would simply mean a return to traditions that were practised in the past, by both European and non-European cultures.

Ulrich Grober has the rare gift of telling a most important story in a clear, accessible and enjoyable fashion. This rich understanding of the many-faceted history of sustainability, in cultures both East and West, provides the inspiration we need to work to reorient our societies towards the same goal. We owe it to our children, our children's children and to all future generations.

Prof. Klaus Bosselmann
University of Auckland
July 2012

CHAPTER ONE

"... an innate ability ..."?

Prologue

Last year's harvest had been meagre. In the villages of the arid regions of western Senegal in 2008, March had already seen the onset of the *soudure*. The French word for a soldering joint, or, by analogy, for any gap which needs to be bridged, has a particular secondary meaning in francophone Africa. It signifies the period between the point when the stocks put aside from the previous year's harvest begin to run out and the start of the new harvest. In the local language the term is *ngekh*. Usually this period of scarcity lasts from early June to mid-September. But in 2008 the rainy season held off. The first fruits of the new harvest didn't appear on the table until late October.

In many regions of Africa, the following scene is repeated every year. After the harvest, every family fills small leather bags with millet, barley or rice. These are laid down, as cool and dry as possible, in the furthest corner of the storehouse. What the farmers are setting aside, are keeping in reserve, are the seeds for the coming year – their life insurance, the only one they have. The little bags lie there, invisible to prying eyes, safe from hungry mouths. Even when the fruits of the last harvest have all been consumed.

Africa is a great teacher. Her ancient wisdom tells us that the human community consists of those who went before us, those who are alive here and now, and those who are yet to come. She teaches us *resilience* – that is, the ability to withstand repeated blows of all kinds, and to mobilise our powers of resistance, in order not just to survive periods of hardship but actively to overcome them, and in so doing to preserve and to strengthen our courage, our zest for life, our cheerfulness. Mia Farrow, the Hollywood icon and Darfur activist, speaks of the "resilience of the soul". We could also speak of *indestructibility*. It is a quality we will be in urgent need of in the future, not just in Africa but all over the world.

In periods of catastrophic drought, such as the 1980s, the *soudure* is unbearably prolonged. The iron rule of survival then is to sell all your worldly goods rather than touch the seeds laid aside. Slaughter your cattle, your goats. Send your children to work in the city. Roam far and wide to earn a little money or food. But keep your seeds. Don't bring out the little bags until the hunger threatens your very lives. And then think long and hard about whether you are going to open them. When families in the Sahel and other African regions begin to consume the seeds put aside for the next sowing, then they are on the edge of the abyss. And so there began in the middle of the 1980s an exodus along the roads and tracks of the Sahel region which many did not survive. Journalists and aid workers reported at that time how women from the tiny farms would bring out their very last little bag and proudly pile up the grains of millet in front of them on the table as if they were diamonds.

In December 2008 I heard Adama Sarr, the young coordinator of a small NGO from an arid region of Senegal, talking to a small audience about the *soudure*. How can the vicious circle of chronic hunger be broken at a time when the first symptoms of climate change are threatening? Sarr reported on what was going on in the twelve villages covered by his network of women farmers, cattle herders and village teachers. How the members protect the baobab, the African tree of life, plant new trees, lay hedges to protect the plantations from the hot winds; how they create microcredits, and teach people how to build cooking stoves to replace the open fires which devour their wood supplies; how they set up composting systems, drill wells, teach reading and writing. As for seedcorn, the group encourages a return to traditional, locally cultivated plants, for example the native varieties of millet. Because the imported commercial seedcorn is usually hybrid – which means it is not germinable, and so useless for sowing.

What is the vision that lies behind the work of this Senegalese farmers' organisation? I found it in the brochure that I took home with me from the lecture: *"accéder à un développement durable"*. To achieve a *sustainable development*.*

Seedcorn should not be ground.[1] Timeless wisdom, and a wonderful metaphor for sustainability. However, the phrase originates not in Africa, but in the writings of Goethe. The great poet, who was also a Minister of a small and impoverished German duchy, drew on his own immediate experience in formulating it. His early years in office were marked by failed harvests and famines afflicting the peasants of Thuringia. Even in the green heart of Germany, the *soudure* at that time was growing more acute with every early summer. In July 1779, for example, whole villages pleaded with the Cammer in Weimar, the ducal treasury, for

---

* Phrases which relate to key definitions of sustainability are shown in italics throughout the book.

tax relief. A begging letter from a village near Jena states that nobody knows where the poor folk can find the seedcorn for the next sowing. As Minister, Goethe was involved with the work of the Cammer. Shocked, he recounted to his lover Charlotte von Stein how he had met a man "who in the misery of hunger had seen his wife expire beside him in the barn" and had had to "scrape out a grave for her with his own hands".

Goethe coined the phrase in his Bildungsroman *Wilhelm Meister's Apprenticeship*. A mysterious Abbé, a member of a secretive 'society of the tower' which the hero of the novel, Wilhelm Meister, becomes involved in, hands him some writing 'containing something of importance'. This 'guide', he is told, deals with the 'education of the artistic faculty'. But its second part deals 'with life'. And this part begins with the words, "Art is long, life is short" and ends with this: "Baked bread is tasty and satisfying for one day. But flour cannot be sown and seed-corn should not be ground."

As Goethe was writing these lines, the ducal forestry department was conducting a huge reafforestation programme. In an ordinance from the Duchess Anna Amalia of 1761, it was decreed that the woods of the duchy should be subject to "a new and sustainable forest management plan".

Over the course of the month of September 2008, the international financial system collapsed. The newspapers printed wide-angle photos of the granite palaces of the banks. In this perspective, financial institutions which had until then appeared as solid as the ancient rock of their architecture seemed as precarious as the Tower of Babel. So-called economic experts who had until then boasted of performing surgery on national economies 'without anaesthetic' fell momentarily silent. "Where is my money?" the documentary maker Michael Moore asked a banker on Wall Street. "I don't know," she replied. During those days British capitalism, according to the *Daily Telegraph*, felt as if it was on its last legs. Billionaires wept on camera. Dazed politicians declared that they had glimpsed 'the abyss'. Overnight, they put together huge bail-out programmes. Then they filled 'economic stimulus packages'. The sums of money that now came into play were beyond comprehension. They exceeded by a large multiple the amount which UN experts had calculated as sufficient to free humanity from hunger.

With the fresh money and securities the banks continued business as usual. And it happened again. The European debt crisis of 2011 followed just the same patterns. "I see our system in the painful process of breaking down", wrote Paul Gilding, the former Greenpeace director. "Our system of economic growth, of ineffective democracy, of overloading planet earth – our system – is eating itself up."

There is no alternative? Really? What about calmly allowing unsustainable structures to collapse, while at the same time gradually strengthening extant sustainable structures, and creating new ones – would that not have been a better strategy for coming out of the crisis stronger? But such a strategy would of course depend on an ability to distinguish with precision between what is sustainable and what is not.

I went hiking in those late summer days of 2008, in the mountain world of the higher Ötztal valley, under an azure blue sky, with the thermometer showing 25°C in the shade. Beyond the last houses of the old Tyrolean mountaineering village of Vent I clambered upwards towards the glacier zone. Ahead of me, at 3,000 metres, lay the dirty grey mouth of the Rofenkarferner glacier. Foaming, milky-green water welled from around the edges of the glacier, then plunged down into the valley, flowing considerably more strongly in the heat of the midday Sun than in the early morning. Like almost all of the world's glaciers, the Rofen is shrinking. In the lush meadows to either side of the stony path, arnica and monkshood, saxifrage and purple gentian were in full bloom. To the south, the view to the main Alpine ridge opened up.

The Texel group of peaks, including Similaun and Hauslabjoch, was within reach. Up there on the ridge, in a gully in the gneiss rocks, on an early summer's day some 5,300 years ago the nameless wanderer we call Ötzi had breathed his last, and had found his presumed final resting-place in the seemingly eternal ice. He was one of us – the first European with whom we have come face to face. The paths he took on his trek from the south over the ridge into the Ötz valley are still there. So are the springs where he drank, and the herbs he used as medicines, for himself and perhaps for others.

It is five days' journey on foot from there to Bolzano. I am standing in front of a refrigerated glass case in the museum on the edge of the old city. Only a pane of glass separates me from the ice man. The mummy from the 4th millennium before the start of our own calendar is surprisingly narrow-shouldered and fine-limbed. His desiccated eyes, their original blue colouring still faintly detectable, are turned upwards. The right hand which swung the axe and drew the bow is now grasping thin air. All around the glass coffin the remnants of his equipment are laid out. Every item reflects his semi-nomadic way of life. Everything is thought through to the last detail and adapted perfectly to his natural environment, his needs and his objectives. The boots, with their bearskin soles, leather uppers and lining of woven lime bark, are ideal for the high mountains. The copper axe is masterfully cast; the yew hunting bow almost the equal of modern sports bows for distance and impact. The construction of the frame of the rucksack is considered by modern manufacturers of camping equipment to be perfect for the carrying of heavy loads. Nine native varieties of wood have been used; exactly the right variety has been selected for each distinct function. The diligence with which the full range of native resources has been utilised, and the elegant simplicity which characterises each of his artefacts, are evidence – bridging thousands of years – of a creative spirit. The man from the glacier – is he the archetypal *homo sustinens*? Does he belong to the long genealogical line of the ancestors of sustainability?

One small detail, discovered only later, disturbs the picture. An arrowhead is buried in Ötzi's left shoulder. This, and not cold, snow and ice, is what killed him. Hunters call it a shoulder shot. It seems he was fleeing. Victim or attacker, hunter or hunted in a bloody tribal feud? The previous week, in Vent, in the crowded dining-room of the town's biggest hotel, I had heard a lecture given by a glacier researcher from Innsbruck. There was considerable evidence from research into climate history, he told us, that a freeze had struck the Alpine region during Ötzi's epoch. He surmised that Ötzi's violent death might be connected with the competition for shrinking pasture lands. Climate wars, over 5,000 years ago? Now, in front of the glass case in the museum, I was struck by a strange vision: his icy coffin thawing as a result of global warming, allowing a silent messenger to emerge from the depths of time into our present day.

In order to grasp the time horizon, it is useful to bear in mind that when Ötzi was alive, Babylon was nothing but a pile of mud huts in the land between the two rivers. But it is possible that at that period already, much further to the east, in the river basin of the Indus, voices were being raised in song to the "all-bearing, firmly grounded, gold-breasted Mother Earth". And also in prayer: "What of thee, O earth, I dig out, let that quickly grow over; let me not hit thy vitals nor thy heart."[2] Can one not read these verses, too, later carried over into the Vedic Hymn to the Earth, as an expression of the idea of sustainability? It was in this sense that Indira Gandhi, then Prime Minister of India, quoted them in 1972 in Stockholm, at the first great United Nations environment conference.

One thing seems clear to me. The idea of sustainability is neither an abstract theory dreamt up by modern technocrats nor a wild fantasy hatched by Wood-stock-generation eco-freaks. It is our earliest, our primordial world cultural her-itage. It was Prince Charles, heir to the British throne, who a few years ago posed the question whether there was not, "deep within our human spirit . . . an innate ability to live sustainably with nature."?[3]

But what is *sustainable*? The 'Dictionary of the German Language' published in 1809 by Joachim Heinrich Campe, Alexander von Humboldt's teacher, defines *Nachhalt* (the root of *nachhaltig*, the German word for 'sustainable') as "that which one holds on to when nothing else holds any longer". That sounds comforting. Like a message in a bottle, from a distant past, for our precarious times. "We are searching for a model output that represents a world system that is: 1. sustainable without sudden and uncontrollable collapse; and 2. capable of satisfying the basic material requirements of all of its people."[4] Another message in a bottle, this one from the famous 1972 report to the Club of Rome on *The Limits to Growth*.

In both cases, sustainability is the antonym to 'collapse'. It denotes that which stands fast, which bears up, which is long-term, resilient. And that means: immune to ecological, economic or social breakdown. What is striking is that the two terms, from such different epochs, are remarkably congruent. They locate 'sustainability' in the basic human need for security.

"In an age", said Christopher G. Weeramantry, former judge at the International Court of Justice, "in which we are denuding the resources of the planet as never before and endangering the very future of humanity, sustainability is the key to human survival. It is the concept which needs to be nourished from every discipline, every culture and every tradition." [5]

To get to the inner meaning of this word, you have to approach it from several angles. It is the aim of this book to contribute to a greater clarity and to a heightened sensibility – for all of us – through an approach based on both linguistics and conceptual history. It tells the story of how intuitive precautionary thinking crystallised over long timespans into a specific concept. How under the umbrella of this concept a semantic field emerged which produced such now familiar terms as *ecology, environment, quality of life* and even *management*. How one word gathered the optimistic dreams and hopes from all epochs of human history and bundled them together into a vision for the future. How ancient survival wisdom was married to modern high-tech innovations. It recounts the slow growth of an idea, and the complex interrelationships to the everyday worlds in which that idea grew. But it also identifies the wrong turnings which were taken on the way. This book invites you to take a step back, in order to take the measure of things from the new perspective; and then to use this new scale to judge for yourself the ideological context, the meaning and the associations of 'sustainability'; to comprehend its gravity – that is, its weight – but also its flexibility.

CHAPTER TWO

Word games

Conceptual confusion

What do you understand by the term 'sustainability'? Is it clear or is it cloudy? Too unwieldy? Too bland? A ray of hope, carrying positive expectations? Or just a bore? Does it free the imagination? Does it clarify? Or does it obscure? However you use it within your own vocabulary, you ought to know what exactly you're talking about.

In recent years complaints about the 'inflated use' of the term, about dilution and confusion of meaning, have developed into a mantra. In my work as a journalist I know people who never utter the word without waggling their fingers in the air to signal quotation marks. It has entered into the language of advertising and political propaganda. A 'sustainable diet', 'sustainable recovery from illness', 'sustainable return on capital', 'sustainable freedom from dandruff' – nothing is impossible. Recently, the 'most sustainable motorway ever' was opened in Switzerland. Anyone for 'sustainable golf'? And what about a 'sustainable Las Vegas'? (Google them!)

What exactly is going on here? There's certainly nothing wrong with 'sustainable health', or a 'sustainable recovery'. Everybody understands what is meant – nothing more than a long-lasting effect. The word functions as a synonym for 'durable' or 'permanent'.

But what is meant when somebody speaks of 'sustainable growth'? Continuous growth of gross national product, or of a business empire, with all the associated ecological and social collateral damage? Or is it the growth of green structures within an economy which may be shrinking because it has abandoned its unsustainable structures? And what is a 'sustainable motorway'? One that runs through ten miles of tunnels to protect the rich suburbs of Zürich from noise and pollution? Or is it perhaps the motorway that a community decides not to build at all?

Sometimes, mental and linguistic sloppiness is to blame. But all too often the fog of confusion is deliberate. It's a strategy sometimes called 'greenwashing', by analogy with Pontius Pilate washing his hands of responsibility, or perhaps with the Cold War expression 'brainwashing'. You can make money out of confusion.

The trick is simple, but not easy to see through, because the word leads a double life. Or rather, it has two usages, two levels of meaning, which exist side by side – a deeper one and a shallow one. The deeper one is a political concept incorporating ecological, economic and social dimensions. In the shallow sense, the word really means nothing more than 'long-lasting'. So far, so good. The game of confusion sets in when the levels overlap. When the discourse in general is non-political, but it is suggested that the full political meaning of the term is being employed. By this means a straightforward expectation of profits over a two- or three-year timescale is transformed into a financial return which is *sustainable* – that is, generated in an ecologically and socially responsible way. Worst of all is when 'sustainability' is turned around against the supposedly exaggerated or unrealistic demands of environmental campaigners: for example, the construction of a coal-fired power station is trumpeted as 'sustainable' because it is cleaner than the old one and preserves jobs. In the same way, 'sustainable forestry' became the battle cry of international timber companies when taking over large areas of tropical rainforest from the indigenous populations.

Once the term has been hijacked and robbed of its substance, there's not much you can do with it. Or rather, nothing you can't do. The most mundane of activities, even the most ruthless pillaging of the planet, can be sold under the hollow label of 'sustainability' or 'sustainable development'. A short scene from 'Alice Through the Looking-Glass', Lewis Carroll's 19th-century children's book, describes the dynamics of all semantic power-games.

> 'When *I* use a word,' Humpty Dumpty said in rather a scornful tone, 'it means just what I choose it to mean -- neither more nor less.'
> 'The question is,' said Alice, 'whether you *can* make words mean so many different things.'
> 'The question is,' said Humpty Dumpty, 'which is to be master – that's all.'
> Alice was too much puzzled to say anything . . .

In the background of that conversation can be heard an ancient voice, that of the Chinese sage Confucius. When he was asked what he considered the first thing to be done if he were to run the government, Confucius answered: to rectify the terms. *Zheng Ming*, as it is called in Chinese – rectifying the names, or terms – is regarded in Chinese philosophy as a starting point for rectifying behaviour.

To reclaim genuine or verifiable sustainability – to defend the concept against abuse and dilution, to champion its strong version as opposed to the shallow and weak versions, and to encourage the reader to adopt this version into their own vocabulary – that is the aim of this book.

A closer look

Sustainability, Nachhaltigkeit, hållbar utveckling, desarrollo sostenible, chi xu fa zhan – the word has spread all over the global village. If you Google it – in only a couple of the language options – then within seconds you get many millions of search results. Taking the internet as a measure, there are not many topics which concern humanity as strongly in this second decade of the 21st century. Anybody engaged with it is surely part of a growing, interconnected, creative global movement, a collective quest. Sustainability, it has often been said, is the key to human survival.

But the term resists definition. This book is not an attempt to give one. My way of approaching the inner meaning, the essence of the term, is to tell its long history.

Our modern concept has surprisingly deep roots and a long, little-known tradition. Old words are usually composed of multiple historical layers. I would like to strip back these archaeological layers in order to get at the core of the word's meaning. This requires a multilingual approach. Let us first cast an eye on the form of the word as used so prominently in the UN's 1987 Brundtland Report: *sustainable development.*

What exactly does *sustainable* mean? One element of the word's morphology is quickly explained: *-able*, derived from the Latin *habilis*, means having the capacity or power to do something. More complex is the verb *sustain*. The Oxford English Dictionary of 1961 (that is, before the new meaning was created) has several columns on the entry and dates it back to the Middle English period (*c.*1150-*c.*1350). The group of meanings which interests us appears under heading 4, where *sustain* means "to keep in being". Another definition is "to cause to continue in a certain state". Then: "to keep or maintain at the proper level or standard". And "to preserve the status of".

Sustain is a word of Latin origin. In the Latin dictionary first published by Lewis and Short in 1879 we find the verb *sustinere*. Its root words are *sub* (under) and *tenere* (to hold). The English equivalents offered are: to hold up, hold upright, uphold, bear up, keep up, support, prop, sustain. This list focuses on the arrangement of objects in space, that is, the supporting function or capacity of a structure. It is followed by definitions which emphasise the temporal dimension, that is, the capacity to continue supporting over a timespan: to hold or keep back, stay, maintain, endure, preserve.

This brings us very close to a technical term which I believe to be the blueprint for our modern concept. It comes from the terminology of forestry. Long before the Brundtland Report, foresters all over the world spoke of 'sustained-yield forestry' when discussing the long-term economic, ecological and social aspects of their work. It was their guiding principle, indeed their holy grail. The term was a translation of a German word: *nachhaltig*. This was introduced 300 years ago. It literally meant: to hold back reserves for future generations. We shall come back to these roots.

A collection of formulations

There is no comprehensive, universal definition of sustainability. The concept is too complex and too dynamic to allow one. Rather, several different formulations are in circulation, all to a greater or lesser degree circumscriptions and/or approximations. Four of these formulations have shaped the discourse to date.

The best-known is a passage from the Brundtland Report published by the UN in 1987: sustainable development is "development that meets the needs of the present without compromising the ability of future generations to meet their own needs".[1] That is the original wording of the globally most frequently-used formulation of the basic idea. Let's call it 'Formulation One'.

The *triangle of sustainability* is a groundbreaking conceptualisation which was widely adopted following the Rio de Janeiro Earth Summit of 1992. Ecology, economy and social justice form the three points of a triangle; they are interconnected, and should therefore be thought of as a network. This is Formulation Two.

Formulation Three is plain and simple: *don't cut down more trees than the forest can re-grow*. This is how foresters express their own classic conception of sustainability. And it offers a way of making the expanded and renewed contemporary conception of it more concrete and easily comprehensible.

Formulation Four: *to preserve creation*, which harks back to the creation story in the Bible and to God's command, *abad* and *shamar* – to tend and keep the Earth. The creation myths of other cultures have very similar commands.

Each of these formulations captures something of the essence of the concept. But as with anything formulaic, there is a dual danger – of diminution and of erosion. When it has been heard and read a hundred times, a phrase loses the power to inspire.

Yet these four guiding phrases can serve as an excellent navigation system for our time journey into the deeper layers of meaning of the concept. This journey will take us to the seemingly harmonious and stable world of the medieval monasteries, the era of the cathedrals (Formulation Four). From there we move on to the geometrically mapped forests of the Enlightenment (Formulation Three), then to the era of the Romantics, the 'back to nature' movement and the discovery of the links between ecology and economy and social equity. And finally back to our own day, to the comprehensive crisis, 'Earth politics', and the great transformation (Formulation One).

Our journey begins in the turbulent and creative years around 1968. Why then? The answer lies in the intellectual and spiritual vacuum of the present. We no longer have a grand narrative, no visions to inspire us and lead us forward. This vacuum is not good. The experts tell us that in order to prevent a climate catastrophe, even at this very late hour, we would need a new 'Apollo Project': an overarching idea which could release huge energies for a common undertaking, no matter what the costs. In the 1960s this was the project to put a man on the Moon, and to bring him back, within a decade. But then something unexpected happened. "We went to explore the Moon and discovered the Earth." These famous words of the astronaut Eugene Cernan expressed the unconscious agenda of the Apollo Project. Just as the delicate, luminous pearl grows in the hard grey shell of the oyster, so a new, Earth-centred, civilising impulse came out of a reversal of perspective experienced by a growth-obsessed, technophilic and expansionist culture. Its *leitmotif* became 'sustainability'. Its foundation was the conviction that the Earth is the most beautiful star in the heavens. This is a promising starting-point.

Creative thinking is nourished by creative imagery. The images – and the icons – of the counter-cultural movements of the Sixties are still with us. Stored in the collective memory of the global village, reproduced millions of times over in cyberspace. With a couple of clicks of the mouse a fascinating little picture gallery can be called up. Earthrise is our first search term.

CHAPTER THREE

"The most beautiful star in the heavens"

Icon Earth

Californian hippies were the first to come up with the idea: show us the whole Earth – just how it looks from space! This request to NASA, spread by a few longhairs on the American west coast via buttons and stickers, was very much in tune with the *zeitgeist*. There, in the stronghold of the countercultural revolution, it was hoped that a photo of the blue planet would have a mind-expanding effect like that of a euphoria-inducing drug. NASA was at first interested in very different kinds of picture – in shots of possible landing sites, and ultimately of the American flag flying on the Moon. But they too changed course very quickly.

In 1968 it finally happened. On Christmas Eve of that year, the human race for the first time saw itself and its planet from the outside. The Earth arose out of the blackness of space over the horizon of a grey, stony, barren moonscape. On Earth, 400,000 kilometres away, the border between night and day is at that moment passing as a semi-circle over Africa, its left-hand margin touching Antarctica. The Atlantic is visible. America, the North Pole and Europe are hidden under thick, swirling cloud. The spacecraft Apollo 8 is orbiting the Moon, about 100 kilometres above the surface. NASA is searching for a suitable spot for the first Moon landing, planned for 1969. A few hours later, still Christmas Eve, Apollo 8 sends television pictures back to Earth. Anders and Lovell, the astronauts, accompany those pictures with a reading from the creation story. "In the beginning God created the heaven and the Earth . . . " They close with the words "and God saw that it was good." Seconds later, their spacecraft disappears behind the dark side of the Moon.

The photo of the Earthrise sparked the imagination of people all over the world. The contours of a new view of the world, of the planet, and of human-kind, began to appear. The American poet Archibald MacLeish, then 77 years old, a veteran of the First World War and a flag-bearer for Roosevelt's New Deal, proclaimed it in the *New York Times*:

"To see the earth as it truly is, small and beautiful in that eternal silence where it floats, is to see ourselves as riders on the earth together, brothers on that bright loveliness in the eternal cold – brothers who know now that they are brothers." [1]

Dante's medieval representation of heaven and hell and of man as the counterpart to God and the centre of the universe was now overthrown, MacLeish argued. But the modern, nihilistic view of human existence as absurd, at the mercy of blind and senseless forces in the margins of a meaningless galaxy – this, too, was now untenable. 'All men become brothers' – *alle Menschen werden Brüder* – echoes of Schiller's 'Ode to Joy' are unmistakable. Yet the ageing poet was also articulating the spirit of 1968 and the euphoric conviction of the dawn of a new age – or 'the dawning of the age of Aquarius', as the 'Hippie musical' *Hair* had it.

The Earthrise photo provided a powerful counter-image to the icon of doom whose nightmarish power had overshadowed the thought and sensibilities of an entire generation: 'Mushroom Cloud', the apocalyptic image of the atom bomb threatening to wipe out human life on Earth. This necrophiliac iconography was created in the period between 1944, when the first atomic explosion took place in the New Mexico desert, and 1963, when an international treaty prohibited nuclear testing above ground. The photos depict the moment after the nuclear explosion. A slim column of steam or dust rises up into the sky from 'Ground Zero', the point of impact of the bomb, then spreads outwards in the atmosphere like the cap of a mushroom before descending back to Earth in the form of deadly nuclear fallout. The dark majesty of these Cold War images threw a long shadow, and they still reverberate subliminally today. One only begins to grasp the liber-ating power of the Earthrise photo when it is set against the doomsday icon which preceded it. And this power was doubled again by a photo which appeared four years later.

7 December, 1972, shortly after midnight local time. In the Kennedy Space Center, the countdown for Apollo 17, still to date the last manned Moon flight, has begun. The launch pad is flooded with dazzling light. It illuminates the snow-white tower of the Saturn V rocket, with the tiny spacecraft in the tip. Wernher von Braun's technological miracle, modernity's answer to the Egyptian pyra-mids. An audience of millions has gathered on the Florida coast to experience

live the grandiose spectacle of light, sound and human audacity – the Woodstock of the space travel-freaks. "Lift off!" A shock wave causes the whole area to quake. In the midst of the inferno of thunder, smoke and flames the rocket rises from the ground, pushes through the fireball it has itself unleashed, and disappears south-eastwards into the dark sky above the Caribbean, trailing a huge cloud of burning kerosene and oxygen.

The perfect moment for the shot arrives very soon, once the spaceship has left Earth orbit and embarked on its elliptical course to the Moon. The men on board, now free of the enormous psychological and physical pressure of take-off, put aside their spacesuits and enter into a state of weightlessness. At that moment of detachment from the Earth, they look back behind them. "Yes, the Moon is there," Evans reports to the control centre back on Earth, four hours, 47 minutes and 45 seconds after the start according to the NASA transcripts. And then, in an ecstatic tone, "The Earth is – that's the Earth . . . Whoo, what a beauty!"[2] A few minutes later, Commander Cernan reports that he is seeing the Earth more full, more complete than anyone has ever seen it. "You know – and there's no strings holding it up either. It's out there all by itself." Harrison Schmitt enthuses about this "fragile-looking piece of blue in space." What has enraptured him and his colleagues is the view of the daylight side of the globe in the full illumination of the Sun. They are far enough into space to be able to see the entire Earth at once, yet still close enough to make out cloud formations and land masses and the slow-motion turning of the Earth. Africa is still fully in view. The eastern tip of South America will shortly appear. This is the moment when the now classic photo is taken. It is probably Harrison Schmitt who presses the button of the on-board camera, a Hasselblad. Five hours after the start, about 45,000 kilometres from the Earth.

On the return flight the spacecraft flies over the desolate waste of the dark side of the Moon. Then the blue planet comes back into view. "You look out of the window and you're looking back across blackness of space a quarter of a million miles away, looking back at the most beautiful star in the heavens. . . You can watch it turn and see . . . it's moving in a blackness that is almost beyond conception."[3]

The landing takes place 302 hours after take-off (59 seconds later than was calculated beforehand), exactly in the target area south-west of the Samoan Islands in the southern Pacific. A capsule three metres across, battered, bruised, blackened. This is all that remains, twelve days after the launch into space. First examinations indicate that the three astronauts are in the best of health and . . . extremely happy. On the 23rd December, just in time for Christmas, NASA releases a photo brought back by Apollo 17. 'Blue marble' became the most widely published photo in media history. What was it that gave this picture its unique aura?

At the moment the shot was taken, the flight path of the spaceship is just crossing the imaginary line joining the Sun and the Earth. The Sun is therefore directly behind the spaceship, so that the daylight side of the Earth is fully illuminated. The Sun's light floods the entire disc of the Earth, almost without any shadowy twilight zones. The planet is a shimmering blue. It is the atmosphere which appears in this colour, where the molecules of air reflect the blue spectrum of the sunlight. A soft blue veil is draped over the land masses. The oceans shine a deeper blue. Large areas of the Indian Ocean and of the southern Atlantic are visible. Giant bands of white clouds swirl around the West Wind Drift. The dynamics of the great wind systems can be seen, suggesting the ultimate forces behind them – the oceanic currents and the power of the Earth's revolution. The South Pole is tilted towards the Sun. The glaciers and ice shelf of Antarctica lie dazzling white in the sunlight, encircled by cyclones. The whole of Africa, the cradle of humanity, is in view, as well as – at the top edge of the globe – the Nile delta and Sinai, the Arabian peninsula and the eastern Mediterranean, centres of early advanced civilisations. Clouds are building up in towers at the Equator, obscuring the surface of the Earth. The green of the tropical rainforest belt shows through only patchily. Areas of high and low pressure alternate. The atmosphere over the Sahara and the Sahel to the north, and over the Kalahari to the south, is cloudless; the warm, earthy colour tones of the deserts, red-yellow-brown, show through strongly. The life-supporting mantle of air around the Earth looks transparent and vanishingly thin, the sleeve of vegetation like soft down. No artefact anywhere which might betray the existence of humankind. Rather, it is the Earth's biosphere which distinguishes it from all other stars, which makes it unique. The blue planet floats through this deep black void of endless space. Her long, slow motion strengthens the impression of dream-like beauty, of absolute solitude and singularity, and – not least – of great vulnerability. Nowhere else in space is there a trace of life. Only one Earth. We are alone.

Fragile is a key term in the contemporary interpretation of these magnificent pictures. The overview photos taken between 1968 and 1972 combined together with the eye-witness reports of the American astronauts and Russian cosmonauts. These men thought of themselves as the outstretched feelers of humanity and interpreted their deeply-felt impressions in almost identical metaphorical language. This quickly coalesced into a grand, compact narrative. It is the story of the immense majesty radiating from the sparkling blue and white jewel. The globe of the Earth, a light delicate sky-blue sphere laced with slowly swirling veils of cloud, rises like a small pearl in a thick sea of black mystery – a sapphire on black velvet. Blue is the colour which results when the sunlight reaches the planetary mantle of air, or – in symbolic terms – when heaven and Earth mix. It is the colour of mysticism, of transcendence. The blackness of space beyond, in its unfathomable depth and unbreakable silence, is deeply unsettling. The stark contrast between the luminosity of the Earth, the black emptiness of space and the cold majesty of the stars emphasises the Earth's uniqueness. But the whole realm of our existence can be encompassed in a single glance – that's how tiny it is. And one glance is enough to reveal that the Earth is not subject to man. From space you can see that all of the Earth's systems are connected with each other. The atmosphere, the land masses and the oceans together with the biosphere make up the space in which life is possible. But of humanity there is not a trace. The human footprint is too small to be visible from up there. "That lonesome, marbled bit of blue with ancient seas and continental rafts is our planet," said Harrison Schmitt, "our home as men travel the solar system." "I embraced the whole planet," writes the astronaut Russell L. Schweickart. His Russian colleague Boris Wolynow adds: "When you look at the Sun, the stars and our planet . . . you get a deeper connection to all living things."

The tone of wonder and of awe, this attitude of humility adopted by people who had shown extreme personal courage in their journeys of discovery into space, is inseparably linked with the photos. "The challenge for all of us," wrote Harrison Schmitt, the man who took the photograph 'Blue Marble', "is to guard and protect that home, together, as people of Earth." In this modern-day saga, the bold adventure, the technological miracle, the aesthetic fascination acquired an ethical and spiritual dimension. Reproduced millions of times over, these pictures became accessible to all of the inhabitants of the developing global village. They required no specialist knowledge of astronomy, geography or ecology. Everybody, even if they couldn't read, could gaze at them, marvel at them, understand them directly and personally. The reversal of perspective produced a sense of common identity no longer limited by geographical proximity but encompassing the whole world.

In 1930, in *Civilization and its Discontents*, Sigmund Freud described the "perception of infinity" as an "oceanic feeling": a feeling "as of something infinite, boundless, oceanic so to speak".[4] Freud thought this exceptional state of the extension of the ego into the world, in which the boundaries of the self are broken down, should be described as 'religious' in a sense which is beyond any systematic religion, beyond faith and beyond illusion.

The two photos from space have been termed icons of our era. The worship of icons is a feature of the Russian Orthodox Church, where an icon is understood to be more than an image, and more too than a symbol. The magical power of the icon allows us to see things which are usually hidden; it opens a window on infinity. It is beyond explanation, touched by the sacred itself. In the cosmic icon we saw for the first time the face of Gaia, of Mother Earth.

The new global narrative, its icons and its keywords – beauty, solitude and fragility – provided the matrix for the discourse on sustainability. The UN's

Brundtland Report appeared 15 years after the last lunar flight. It starts with these words:

> In the middle of the 20th century, we saw our planet from space for the first time. Historians may eventually find that this vision had a greater impact on thought than did the Copernican revolution of the 16th century, which upset the human self-image by revealing that the Earth is not the centre of the universe. From space, we see a small and fragile ball dominated not by human activity and edifice but by a pattern of clouds, oceans, greenery, and soils. Humanity's inability to fit its activities into that pattern is changing planetary systems, fundamentally. Many such changes are accompanied by life-threatening hazards. This new reality, from which there is no escape, must be recognised – and managed.[5]

Silent Spring

Silent Spring. A powerful symbol, and the title of a book by the American marine biologist Rachel Carson. Her topic was the war of Man against Nature. The book appeared in 1962 and contributed to the rise of environmental awareness around the world. Silent spring – the reference is to a delicate sound pattern: the rising and falling cadences of the song of the robin, with its striking range and melodic and romantic quality. The trilling and piping of the little songbird, which can be heard both early in the morning and late in the evening, from the cock but also from the female hen, has a particularly uplifting effect on the human psyche. So this metaphor implicitly asks what we lose if – by way of example – the song of the robin is extirpated from our environment. So it is possible to argue that the modern environment movement was set in motion by the threatened extinction of natural beauty.

"There was once a town in the heart of America where all life seemed to live in harmony with its surroundings."[6] This is how the 'Fable for Tomorrow' in the opening chapter of Rachel Carson's book begins. "The town lay in the midst of a checkerboard of prosperous farms, with fields of grain and hillsides of orchards where, in spring, white clouds of bloom drifted above the green fields". . . A picture of a safe and secure world; but then comes the shock. "[A] strange blight crept over the area and everything began to change." The next spring came. "There was a strange stillness. . . The few birds seen anywhere were moribund; they trembled violently and could not fly. It was a spring without voices. . . only silence lay over the fields and woods and marsh."

The idyll is "contaminated". The original Latin meaning of the word Rachel Carson uses here means not just soiled or poisoned, but also stained, besmirched, profaned. *Silent Spring* is the first frontal attack on the 'elixirs of death' of the chemicals industry, which from the middle of the 1950s onwards were sprayed from small planes over wide areas of many parts of the USA – to 'combat pests', for the 'protection' of plants. It was above all women who contacted Carson, then already a famous nature writer, towards the end of the 1950s. One housewife from the Midwest wrote to her that following a DDT-spraying campaign all the local robins had disappeared within a year. A journalist from a small town on the east coast told her how she had found seven songbirds lying dead at her birdbath after one such sortie. A Waldorf schoolteacher from New York, who owned a Long Island garden in which she cultivated herbs and vegetables on biodynamic principles, sent her the legal papers from a case she had pursued – ultimately without success – right up to the Supreme Court following the contamination of her garden.

What was the background? Around 1954, the authorities began to deploy the chemical agent DDT on a massive scale in order to check the ravages of Dutch elm disease, then rampant in North American parks and avenues. But the leaves of the trees treated by this means fed the worms in the autumn; the poison was laid down in the tissue of the worms; and the worms were eaten, or fed to their chicks, by the returning songbirds in the spring. Food chains became death traps. A spring without the dawn chorus of the songbirds in the woods, in the fields or in our gardens. Is this something we want? Do we really want this sacrifice? Is this natural phenomenon not a constituent element of our emotional wellbeing, and thus of our prosperity? It was this question which Rachel Carson made the starting point of her book.

With masses of supporting facts and arguments she documents the devastating impact of DDT and other chemicals on life in the earth, in the water and in the air. The author denounces with equal anger – at the time she wrote the book she was already terminally ill with cancer – radiation from Strontium-90, present in the fallout from nuclear testing. Her message: 'Nature will fight back' – unless we urgently rethink, stop contaminating our world, and return to the principles which nature itself employs. Rachel Carson's book freed the emergent, little-known science of ecology from its ivory tower and popularised its essential message: in nature, everything is related to everything else.

Following the publication of *Silent Spring*, Rachel Carson was subjected to an unprecedented campaign of defamation in the media, directed by the chemicals industry, in which she was denounced as hysterical, a liar, a fraud and a communist. She had to endure questions about why a childless woman should concern herself with issues of genetic inheritance. Rachel Carson died in the spring of 1964. Eight years later, in 1972, the US representative at the UN's Stockholm Environment Summit announced a ban on the use of DDT in agriculture. Ecology had become the driving concern of a worldwide movement. The fascinating view of the blue planet was now linked with a new concern for the immediate environment, an awareness that the exquisite sound patterns of the robin audible in our gardens had not been found anywhere else in the entire known universe.

Rachel Carson approached the concept of sustainability from the perspective of ecology. She writes of the conservation of the Earth for future generations. All life must seek a state of adaptation and balance with its surroundings. This was her lifelong theme. Already in 1953 she had denounced in a letter to the *Washington Post* the reckless exploitation of offshore oil deposits: "The real wealth of the Nation lies in the resources of the Earth – soil, water, forest, minerals, and wildlife. To utilise them for present needs while insuring their preservation for future generations requires a delicately balanced and continuing program, based on the most extensive research." The essence of the Brundtland formulation, expressed by a prominent ecologist and journalist more than three decades earlier.

Rachel Carson gave *Silent Spring* a dedication: "To Albert Schweitzer, who said: 'Man has lost the capacity to foresee and to forestall. He will end by destroying the earth.'" This quotation contains a wonderful implied definition of sustainability: it is *the capacity to foresee and to forestall*.

In August 1965, three years after *Silent Spring* appeared, a small scientific conference which attracted little attention at the time took place in the university city of Boulder in the Rocky Mountains. Its title was 'Causes of Climate Change'.[7] The experts gathered there were of the view that the planetary climate was a precarious system, susceptible to human intervention and with a dangerous capacity for dramatic change. The conference was presided over by Roger Revelle, like Rachel Carson a marine scientist. The first scientific data-gathering he had been involved in was for the US Navy, in the aftermath of the nuclear tests on Bikini atoll in the Pacific. One year after the Boulder conference, in 1966, Revelle gave a lecture at Harvard University. Sitting in the lecture theatre was an 18-year-old student from Tennessee named Al Gore.

Four years after that, in the summer of 1970, in the middle of the short era of Moon flights, the youth of America celebrated *Earth Day* in its city parks and in its wild open spaces. 20 million people protested against the destruction of their environment.

In Germany, the influential magazine *Der Spiegel* had kindled public interest in the new topic in November 1969. The title of its story was 'Apocalypse 1979 – Man is poisoning his environment'. The stone had begun to move.

The cry of the butterfly

At that time, new ideas were imbibed not just through books, but more quickly and more intensively through a politically conscious and unbelievably creative pop culture. Young musicians instinctively poeticised, radicalised and popularised the questions being asked by the new movements in their own medium. "Well, take me back down where cool water flows . . . Walking along the river road at night, / barefoot girls dancing in the moonlight – green river . . ." In the summer of 1969, only a couple of weeks after the first Moon landing, after Woodstock, I heard that song on the radio. 2:14 – a couple of repeating guitar riffs, an evenly pulsing bass line, the driving drums, the raw voice of a 24-year-old telling us about the happy childhood summers he thought would never end, about days and nights at the green river: learning to swim, the cries of the bullfrogs at night, barefoot dancing girls in the moonlight. 'Green River' was a hymn to untouched nature and to the human communities which lived in and with it.

One year later, in August 1970, I arrived on the West Coast, having hitch-hiked, over the course of several weeks, right across the USA. The next evening, together with the folks who had taken me in, I drove over the Golden Gate Bridge from Berkeley to San Francisco. In the warm summer night we heard a live performance of Green River from a distant stage. If I remember correctly, Creedence Clearwater Revival were playing a benefit concert in aid of a newly-founded free drug clinic, and I listened rapt to their song. ". . . you're gonna find the world is smould'ring / and if you get lost, come on home to green river."

I experienced the highpoint of my time as a rock fan two years earlier. I had arrived – hitch-hiking again – in Frankfurt am Main to enrol at the university for the winter semester 1968 / 69. That evening The Doors were playing. The concert, in the Kongresshalle, was short and tumultuous. After the provocative song dedicated to the 'Unknown Soldier', with a dramatic staging of a firing-squad accompanied by drum rolls, a few of the many American GIs in the audience angrily clamoured for 'Light My Fire'. As the organ intro began, a small group of soldiers forced their way onto the stage and waved their regimental banner. Jim Morrison grabbed it from out of their hands, crumpled it up and hurled it into the audience with an obscene gesture. That was it. Immediately the song was finished, the lights came on. The Doors left the stage, the audience left the concert hall. I remained behind. As I still didn't have anywhere to sleep and had to leave early the next morning I wanted to stay as long as I could in the warmth. The lighting crew and the cleaners went about their work in the almost empty auditorium. But suddenly The Doors came back on stage, picked up their instruments and started jamming.

"When the music's over." In the middle of the piece comes a gentle guitar section, a calm bass line, a restrained vocal: "Before I sink into the big sleep / I want to hear the scream of the butterfly." Unlike in Rachel Carson's book, the subject here is not the song of the robin, but the surreal scream of the butterfly. "I hear a very gentle sound, very soft, very clear." And then loud and clear the spoken passage of the song: "What have they done to the Earth?" On the almost pitch-black stage, Morrison, now softly, almost silently, now screaming, evokes images of the rape of the Earth. "What have they done to our fair sister? / Ravaged and plundered and ripped her and bit her / Stuck her with knives in the side of the dawn / And tied her with fences and dragged her down." Then, quietly again: "I hear a very gentle sound." And then, in a primal scream: "We want the world and we want it . . . now. Now? Now!" I had goosebumps all over.

The hierarchy of needs

In 1970, a new edition of Abraham Maslow's standard work of psychology *Motivation and Personality* appeared in the USA.[8] The book had already been well-known within a small circle in the 1950s, but now it captured the *zeitgeist*.

Maslow's theory of basic needs is often depicted as a pyramid. It consists of five levels representing a fluid ascending hierarchy of needs. The basic level is made up of the essential bodily needs to ensure survival: food, water and air, a roof over one's head, satisfaction of sexual needs. Once these basic needs are adequately met and there is good reason to believe that they can be adequately met in the future, then they recede from the forefront of the individual's consciousness. They are replaced there by a new set of needs: the pursuit of safety and security, bodily integrity, stability and freedom from fear. Then comes the level where the successful social integration of the individual is the dominant concern. What is important here is a network of loving human relationships, of affection, of 'belongingness' within a cohesive community. After this, the desire to develop a stable self-esteem makes itself felt. The respect and good opinion of others come to the fore. And respect which one has actually earned is especially important for grounded self-respect. The need for beauty and the pursuit of knowledge and consciousness come into play.

At the apex of the pyramid is the need for 'self-actualisation'. This becomes a major motivation for feelings, thoughts and actions, and can even push all other needs into the background. It derives from the deep human drive for a fulfilled life. Maslow defines the meaning of this new term by quoting Nietzsche's aphorism, "Become what thou art!" – meaning develop to the fullest extent of which you are capable. "A musician must make music, an artist must paint, a poet must write, if he is to be ultimately at peace with himself." Self-actualisers, he says, are people who observe themselves and their environment, their strengths and weaknesses, with precision and with fundamental acceptance. They possess inner autonomy and a stable sense of community as well as great openness and a capacity for life and enjoying it. They are free of unnecessary neurotic feelings of shame and guilt, and not dependent on the approval of others. However, they are capable of deep connections to other people and to nature. The contradiction between selfishness and selflessness is resolved in such people. Self-actualisers dedicate their lives to the complete development and realisation of their aptitudes, possibilities and potential. Their behaviour is marked by "spontaneity, simplicity and naturalness". They are not frightened by the unknown. They meet and appreciate strangers and new experiences with curiosity, wonder and reverence. They are problem-focused and interested in solutions. The mysterious, the new or the strange attracts and excites them. They follow their inner voice and a vocation which usually emanates from outside, from society. The pursuit of excellence, of peak experiences and of the sensation of oneness with nature and the cosmos are substantial elements of self-actualisation.

The concept was not new. Maslow had found it in the work of the neurologist and psychologist Kurt Goldstein, a German émigré to the USA in 1933. But the basic idea was already present in the ideal of *Bildung*, of education, of the German classical period. Wilhelm von Humboldt provided the formulation for it: "The development of all the seeds . . . which lie within the individual make-up of a human life is what I take to be the true purpose of human existence."

In a conscious turn away from the classical schools of psychotherapy, Maslow focused on the factors which make for psychic health and a successful life. With this in mind he sought out trial subjects. In the Native American lands, he interviewed people who still lived according to the traditional value system of their tribe. He studied the biographies of forceful personalities, among them his contemporaries Eleanor Roosevelt, Aldous Huxley and Albert Schweitzer, and historical figures such as Spinoza and Goethe.

All five levels of the Maslow pyramid include needs which are felt by everyone. An ascending line leads to the 'metaneeds'. It might not be possible for everyone to reach the level of self-actualisation, but for the achievement of a 'good society' it was of decisive importance that everyone should have the same opportunity to progress to the next level. This is the dynamic principle in Maslow's hierarchy of needs. He thought that within every human being, within the human spirit, there exists a potentiality for inner growth, for the realisation of all innate abilities, for peak experiences. Only this kind of growth 'beyond oneself' makes a rich inner life and deeper happiness possible. So the good life depends not on the satisfaction of basic biological needs by means of an ever-growing quantity of material goods/commodities. It is immaterial goods, activities, objectives, which give meaning and are the crucial factor. The fight against poverty is no longer just the struggle for more. It focuses on *access* to the fullness of life, open to everyone.

In the early 1970s Maslow's theory of needs was linked to the growing environmental consciousness. This combination brought forth a new political and cultural concept which has since then been a fixed component of the sustainability discourse: *quality of life.*

The Brundtland Report picked up on this new thinking. When in that report the needs of the present generation are weighed against those of future generations, what is meant is always Maslow's basic needs, and certainly not the 'needs' of the affluent society. "Sustainable development requires that the basic needs of all are met and that the possibility of fulfilling the dream of a better life is extended to all."[9]

Power to the imagination

John Lennon and Yoko Ono's song 'Imagine' was written in the summer of 1971, in the couple's country house, Tittenhurst Park, near Ascot, on the western fringes of London. They wrote the song in this aristocratic setting, and also shot the accompanying, celebratory three-minute video there. You can still watch it today with one click of the mouse.

As the piano sounds the opening notes we can see both of them among the trees in a foggy park. The camera shows them from the back. A young couple, dressed in black, shoulder to shoulder, hand in hand. "Imagine there's no heaven/It's easy if you try . . ." As the voice sets in, they come out into the open and walk towards the snow-white façade of a classical Georgian mansion. "No hell below us/Above us only sky . . ." Lennon puts an arm around his wife's shoulder. They reach the porch, with its two columns, and pause for a moment. A sign comes into view: 'This is not here'. I understand this to mean that we are not entering an aristocratic mansion but a region of our consciousness: the imagination, the inner palace where the dreams of a better world originate. "Imagine all the people/Living for today . . ." the voice sings as the two people disappear through the entrance. A darkened, semi-circular room. The only light falls through the gaps between the curtains. "Imagine there's no countries/It isn't hard to do . . ." Yoko Ono, now in a white ankle-length dress, opens the first curtain and lets in daylight. "Nothing to kill or die for/And no religion too . . ." The shot cuts away to John Lennon, seated at a white Steinway piano. "Imagine all the people/Living life in peace . . ." Moving from left to right along the row of windows, Yoko Ono opens the curtains one after another, floods the room with sunlight. "Imagine no possessions/I wonder if you can . . ." Lennon's solemn face comes into the picture close up. His gaze shifts between the piano keys and the camera lens. "No need for greed or hunger/A brotherhood of man . . ." The

33

fresh green of the garden penetrates the room with the light, while Yoko strides through the by now bright room and sits down next to her husband. "Imagine all the people/Sharing all the world . . ." The couple can now be seen together behind the piano. A white headband holds back her long jet-black hair. Her gaze is focused inward as he begins the final verse: "'You may say I'm a dreamer/But I'm not the only one . . ." They look into each other's eyes. "I hope someday you'll join us/And the world will live as one." End of the song. Outside a bird is singing. Their eyes shine. A long kiss. The video ends at 3:14.

'Imagine' served as the hymn for a generation. Why? One reason was the melody, played *sostenuto* on the piano and supported by a simple but stylish string arrangement, and sung by Lennon's powerful voice. The composer himself referred apologetically to the "sugar-coating" of the song. The catchy, pushy tune combined organically with the revolutionary message in the lyric. "An anti-religious, anti-nationalist, anti-conformist, anti-capitalist song," Lennon said in an interview. "Virtually the Communist manifesto." Above all, though, the song picked up on the *zeitgeist*. It expresses the revolutionary belief that another world is possible.

The future is never the linear continuation of the present. It is always created anew – using the intellectual and spiritual resources of the past. Every shimmering pearl grows within a hard, grey oyster shell. The new can only develop in the womb of the old. But it is a continual source of wonder that in every renaissance, in every era of transformation, the images, the basic concepts and the vocabulary of sustainability appear in new and different forms.

CHAPTER FOUR

Ur-texts

Canticle of the Sun

Laudato si, mi Signore, cum tucte le Tue creature, spetialmente messor lo frate Sole, lo qual è iorno, et allumini noi per lui.

Be praised, my Lord, with all your creatures, especially through my lord Brother Sun, who brings the day; and you give light through him.

This pious hymn of praise transports us to the world of the medieval monasteries and the era of cathedral building. More precisely, into the parallel world of hermitages on sun-flooded hilltops, of the endless dusty country roads of central Italy, of village poverty and barefoot prophets. *Canticum Solis*, Francis of Assisi's Song of the Sun, the medieval ode to creation, is more powerfully present in our cultural consciousness than any other text from this period. Its 'ecological' message has often been remarked upon. But the Song of the Sun also contains the basic conceptual framework of sustainability. In order to recognise the contemporary discourse in this ancient inheritance we need to go to the sources.

Assisi lies on a hill. When one approaches on foot, the city rises up dramatically ahead: the massive monastery buildings on the westerly spur, the shimmering pink house façades of the old town, the ruins of the castle on the hilltop, the steep, wooded slopes of the Monte Subasio massif to the east. The Umbrian city in the green heart of Italy is a spiritual centre of old Europe.

The *genius loci* is especially potent at two locations outside the city walls. Going out through the Porta dei Cappucini, the highest city gate, one finds oneself on a small road lined by cypresses, then on a woodland path which climbs up the back of Monte Subasio. Up there, between two rocky slopes, lies the hermitage of Santuario delle Carceri. The place served Francis as a retreat. A stone bed with a wooden headrest, on which he reputedly slept, remains in the room.

In the holm oak forest higher up one stumbles upon grottos where he fasted and meditated. The old oak tree by the stone bridge nearby is where legend has it he preached to the birds.

The other magical place is the simple monastery building of San Damiano, surrounded by olive groves and cypresses, on the southern slope of the hill dominated by Assisi. The building, which had been dilapidated, served Francis and his small band as a first accommodation. With his own hands he repaired it and then handed it over to Clare, his follower and companion. There, in the garden of her monastery, Clare built for him a tent "of bushes and rose branches" when in the autumn of 1225, at the age of nearly 50, his strength fading and tormented by pains in his eyes, he wished to come to rest. One year before his death Francis experiences in San Damiano a "shining sunrise in the soul", as the French Franciscan Eloi Leclerc describes it.[1] In a state of ecstasy Francis composes his *Laudes creaturarum*, the Song of the Sun. Codex 338 in the city library of Assisi, dated 1279, is the oldest surviving manuscript. Its language is the 'volgare', the early Italian vernacular, already distinctly different from its Latin source.

The 50 lines of verse take the 'Most High' as their starting-point: "Praise be to Thee, my Lord, with all Thy creatures [*cum tucte le Tue creature*]". The word 'tucte' is revealing. The elevation of the soul does not take place via the denigration of the material world. On the contrary. The soul opens itself up to all creatures, to the whole creation. From the opening, the *Laudes creaturarum* celebrate the fullness, the wholeness, the unity and, repeatedly, the beauty of apparently inanimate matter and of living nature. *Tucte le Tue creature* – in the language of ecology and of Earth systems analysis today: the web of life.

The perspective embodied in the text runs vertically. The arrangement of the images leads from the very top to the very bottom. From the Most High by way of the Sun, the stars and the Moon it cuts through the Earth's atmosphere and reaches the biosphere, the waters and the land. 750 years before the iconic Blue Marble photo, this cosmic song of praise envisages the view onto the planet from above.

But Francis doesn't speak simply of the Sun, the Moon, wind, water, fire . . . He always refers to *frate sole, sora luna, frate vento, sora aqua, frate focu* . . . . Everything is brother or sister. Human beings and natural phenomena come from the same source and have equal rank. They are the progeny of a common father. Something important is going on here: the Franciscan perspective dissolves the dividing line between human beings and the rest of creation. It marks a radical break with powerful traditions of classical and Christian thought – and presents Western modernity with an equally radical challenge. The subjugation of nature was, and is, legitimate in the mainstream tradition; indeed, it was a commandment. It is part of our normality. In contrast, the new human being of a Franciscan, solar civilisation accepts and celebrates being part of nature. This reconciliation provides the spiritual basis for the 'communion', for a universal community of human beings and fellow creatures.

The Sun is singled out for special praise. It is not simply *frate*. As the endless source of daylight and energy, it is simultaneously *messor* – lord. The Sun is accorded particular aesthetic properties: it is beautiful, radiant, brilliant. It is the source of joy and of aesthetic pleasure. It is even, as in so many of the world's cultures, a symbol of the divinity itself. Sun and Moon complement each other. Like day and night, light and dark, clarity and mystery. *Sora luna e le stelle*, the Moon and the stars, are still part of the heavenly realm. In the blackness of the cosmos the twinkling of the stars appears "precious and beautiful" (*pretiose et belle*); the Moon, with its cycle, its gentle energy, seems especially mysterious and attractive.

In the following verse the imagery of the text enters the sphere of the four elements: air, water, fire and earth. "Be praised, my Lord, through Brother Wind." Assigned to him are the air, the clouds, "fair and every kind of weather". So this is about the mantle of air surrounding the Earth, and the differing forms it takes – that is, about the climate. Precisely at this point, the original word for our term 'sustainability', the *Ur*-text, appears in the Canticle of the Sun for the very first time: *sustentamento*. Francis praises God for the phenomena of the atmosphere, through which He gives His creatures *sustentamento*, that is, support, upkeep, sustenance: "*per lo quale, a le Tue creature dài sustentamento*". The term denotes everything which is necessary for the support and survival of living and non-living things: the means of survival, the basic necessities of existence. The provision of these things on a lasting basis is a gift of God. He bestows them not through Brother Wind alone. Sister Water (characterised as very useful, humble and precious) and Brother Fire (beautiful, congenial, robust and strong) play an equal part.

The Canticle of the Sun now becomes a canticle of the Earth. "Be praised, my Lord, through our sister Mother Earth." Like the Sun, the Earth, too – and by that is meant principally at this point the soil, the humus, the native ground – is accorded special status, twice over. *Sora nostra mater terra*. She is not only, as in the Doors song from 1967, our fair sister. Nor is she only *magna mater*, the Great Mother, as in archaic cults, or *Gaia*, the Earth goddess. According to Eloi Leclerc (a French Franciscan and acknowledged expert on the Canticle), the maternal Earth receives in Francis's work in addition the "face of a sister", and thereby a new – eternal – youthfulness. "The feeling of dependence and reverence which is due towards a mother is here nuanced by the feeling of sibling affection." For the Earth is also a sister, and thus daughter by a common father, herself a part of creation.

To be sure, in her capacity as mother she has a special power. *Mater Terra, la quale ne sustenta et governa et produce diversi fructi con coloriti flori et herba*. She is the Earth, who sustains and governs us and produces many different fruits with colourful flowers and herbs. Here, Francis uses for the second time a form of *sustentare*. What sustains us? It is God's Earth, in interaction with the atmosphere around the planet. Today we call it the biosphere. It never fails to bring forth fruit, fruitfulness, biodiversity, and – linked to this last – colour and beauty. As long as we allow ourselves to be *governed* by it. The imagery of abundance, fullness and variety is inseparable from the terminology of *sustentamento*.

It is only from this vantage point that we can understand the core of the Franciscan worldview – its ideal of poverty. The pleasure at the abundance of life all around provides the impulse to reduce the 'consumption' of 'resources' to a minimum. Possessions are a burden; self-denial is liberation. "Take nothing for your journey" is the command in the New Testament [4, Luke 9: 3, KJV]. "Imagine no possessions", sang John Lennon. If one elevates the lack of possessions to a guiding principle, one has to know what will instead provide the security that possessions normally bring. The ancient text tells us that it is nature, if treated like a sister, that can sustain us over time. Confident of this secure foundation we can create a new picture of the good life. Franciscan minimalism is a way of preserving, and so of experiencing and judiciously enjoying, the integrity of creation – of all creatures – their beauty, their solidity, their multicoloured variety. Self-denial, not denial – prohibition – by others, opens the way to the splendid abundance of life. The Franciscan pyramid of needs: to live simply, in equality, in harmony with creation, open to the voice of mystery. To be sure, in this medieval source sustainability is not created by man, but granted to him. It is a gift of God's grace. The foundation of Franciscan theology is the belief in divine providence.

The last word of the Canticle of the Sun is *humilitate* – humility. Francis's last wish was to die lying naked on the bare earth. Legend has it that this is how he met Sister Death. The place where he died, the simple Capella del Transito in the wooded valley below Assisi, was built over in the 17th century with a monumental baroque arched basilica. At just this time, north of the Alps, the French Enlightenment sage Descartes was searching for a "practical philosophy" and "knowledge which would be very useful for life". Like Francis, Descartes concerned himself with the elements. Through his philosophy he wanted to "know the power and action of fire, water, air, the stars, the heavens and all the other bodies in our environment" in order to use this knowledge for all "appropriate purposes". His idea was "to render ourselves . . . masters and possessors of nature". It is difficult to imagine a starker contrast to the Franciscan ideal.

Taking care of Creation

"Man has lost the capacity to foresee and to forestall. He will end by destroying the earth."[2] This was Albert Schweitzer's conclusion. It assumes that Man is potentially capable of saving the Earth. This idea was unknown to medieval thinking. The preservation of Creation was the business of 'divine Providence'. Man's role in this was subordinate. For more than a thousand years, from St Augustine to Martin Luther, the doctrine of the 'Providentia Dei' was a vital part of the foundation of Christian belief. This technical theological term combined a whole range of meanings: predestination, foresight, provision, care. The era of the Enlightenment saw the collapse of belief in Providence. However – this is my thesis – important structural elements of the modern discourse of sustainability were formed out of its ruins. A look at the historical edifice may therefore be very instructive.

The idea can be found already in ancient philosophy: the remarkable purposefulness and beauty, both in the smallest things and in the greatest structures in nature and in the cosmos, point to an organising power. In view of the comprehensive nature of this power it can only be of divine origin. The Greek philosophers Anaxagoras, Plato and Epicurus spoke in this context already of divine 'providence' and 'care' (*tou theou pronoia*). The Stoic philosophers derived their cosmology from the same observation: "All that is from the gods is full of providence. . . . From thence all things flow" wrote Marcus Aurelius, the philosopher and Roman emperor.[3] "All things are implicated with one another, and the bond is holy. . . . For there is one world (*unus mundus*) made up of all things, and one god who pervades all things, and one substance, and one law, one common reason in all intelligent animals." One world – *unus mundus* – then, already! Human beings, gifted with the power of reason, are able to perceive the given universe as meaningful, to enjoy its beauty and to find their place within it. Note: to find their place within it. Not to shape it and 'manage' it on utilitarian principles.

In early Christian theology, the doctrine of Providence is directly connected with the Creation story. After the act of Creation (*creatio*) "out of nothing" (*ex nihilo*), God did not simply abandon his work. His will remains present in what has been created. He continues to work within it by virtue of his omnipotence and according to his plan. 'Providentia' means the *conservatio*, the preservation of the world and the continuation of creation. God cares for all of his creatures, keeps them in existence, and leads them towards their destiny, their *telos*, namely salvation. A document from the very first Christian community of Rome, the so-called First Epistle of Clement, written about 100 AD, concerns itself with the 'sustainability' of the divine will:

> The heavens are moved by His direction and obey Him in peace. Day and night accomplish the course assigned to them by Him, without hindrance one to another. The Sun and the Moon and the dancing stars according to His appointment circle in harmony within the bounds assigned to them, without any swerving aside. The Earth, bearing fruit in fulfilment of His will at her proper seasons, putteth forth the

food that supply abundantly both men and beasts and all living things which are thereupon, making no dissension, neither altering anything which He hath decreed.[4]

The work of Providence, however, exceeds all human experience. Consequently, it is not always transparent. It is for just such cases that St Augustine, a founding father of the Church, brings the "invisible hand" of God into play. God, he writes in his book *De civitate Dei*, is not like those artisans "who use their hands, and material furnished to them"; rather, "God's hand is God's power; and He, working invisibly, effects visible results." This is another metaphor which has had a continued existence and impact in secularised form: Adam Smith had been excavating in the ruins of the doctrine of Providence when in 1776 he sanctified the "invisible hand" of the market.

One thing is certain: a cherry stone will not grow into an apple tree, but always and only into a cherry tree. For the medieval theologian Thomas Aquinas, the dependability and constancy of the course of Nature represent evidence of her teleological or purposeful structure. Perhaps the most influential of the Scholastics, born in 1225, the same year in which the Canticle of the Sun was written, Thomas sees in this reliability the operation of a purposive 'direction'. "Therefore," he deduces, "there must of necessity exist a being through whose foresight (*providentia*) the world is governed (*mundus gubernetur*)." We have come upon the word *governa* already, in the Canticle of the Sun. For Thomas Aquinas, *gubernatio* (steering, leading, direction) becomes a key term. God steers the particular and the whole to the realisation of all immanent potentialities. "The construction of reality", the German theologian Udo Krolzik comments, "is therefore founded on the striving of natural things to become that which they are by virtue of their own nature." *Conservatio* is not inertia, not a static preservation. Growth and development are a part of it. It is *gubernatio* which gives this process a purposeful dynamic.

The doctrine of *providentia* enjoyed a final flowering under Luther. As "everything in heaven and on Earth" was ordered "so wonderfully, beautifully and certainly," Martin Luther preached in 1537, there must be "a single, everlasting divine being who creates, sustains and rules all things". Luther believed in the ubiquity, the omnipresence of the divine, though admittedly the divine presence had "burrowed deep into all creatures and hidden itself away". This view became a powerful spur for research into natural history. The Swedish naturalist Linné, son of a Lutheran country pastor, was still looking 200 years later for the 'footprint of God' in nature. Luther's God is 'hidden'. His Providence is an act of grace.

> Commit thy Ways and Goings
> And all that grieves thy soul,
> To Him, whose wisest Doings

Rule all without Control.
He makes the Times and Seasons
Revolve from Year to Year
And knows Ways, Means, and Reasons
When help shall best appear.

This is John Wesley's well-known translation from 1739 of the chorale "Befiehl du deine Wege", written in Berlin by the Lutheran preacher Paul Gerhardt in 1653 and incorporated a few decades later by Johann Sebastian Bach into his St Matthew Passion. It contains in condensed form the Lutheran doctrine of Providence. It is the same message as that proclaimed in the black Protestant spiritual of the American South, "He's got the whole world in His hands."

Conservatio est actio Dei externa, qua ex mera bonitate omnia, quae sunt, sustentat. "Preservation is the externally-directed act of God which out of pure benevolence sustains all things which are." The words *conservatio* and *sustentare* – an echo of the Canticle of the Sun, a prefiguration of the sustainability discourse of the 20th century – appear in a theological treatise of 1682. Its author, Abraham Calov, was regarded as the the guardian of Lutheran orthodoxy at the university of Wittenberg. His authority was widely recognised throughout northern Europe. Calov was the *spiritus rector* of the pious Paul Gerhardt, and a contemporary of Kepler and Descartes. On the threshold of the Enlightenment, the Christian doctrine of *providentia* appears here as a highly sophisticated system, with all of its conceptual framework of supporting columns, braces and adjusting screws fully visible.

The guiding concept of *providentia* signifies the action and the capacity of thinking ahead into the future. Anticipation leads to knowledge of things in advance. What has been seen in advance and corresponds to what is willed requires the agency of caution or attentiveness to bring it about – that is, of precaution. This in turn demands in any given moment an action (*actio*) which is tailored to the desired end. The structure of Providential action is highly complex. It has three fundamental elements. The most basic is *conservatio* (synonymous with *sustentatio*, the Franciscan *sustentamento*). This refers to the conservation, the preservation of all things in the existential consummation settled upon them by the Creation. It is a continuation of the Creation, which prevents a lapse back into nothingness – *annihilatio* or annihilation.

Gubernatio means the guiding and steering of all processes, the governance over things. One element of this is *cura*, care and ministration, but also the curation or stewardship of creation. The fixed natural phenomena are evidence of divine *gubernatio*: the regularity of the motion of the heavenly bodies, the cycle of the seasons, the water cycle, the empirical rules of the science of genetics.

The third element of *providentia* is *concursus*, the combination or interaction between several causal principles. This relates to the relationship between divine action (*actio externa*), the operation of natural forces and human free will. Divine action is *prima causa*, the first or fundamental cause. The cooperation of natural forces and human action (*cooperatio*) are *secundae causae*, second-order causes, though with their own scope, effects and side-effects. Finally, the doctrine of *concursus* also addresses the ticklish question of the role of evil in the course of events. Divine *gubernatio* shifts between different options: permission (*permissio*), obstruction (*impeditio*), alignment with divine aims (*directio*), and limitation (*terminatio*) of evil.

This elaborate structure of ideas stood for over 1,500 years. Then the times changed, and the building collapsed. The key dates in this disintegration were 1666, when Newton observed in his parents' garden how an apple fell to earth, and 1755, when an earthquake destroyed Lisbon and dragged tens of thousands to their doom, indifferent as to whether they were believers or heretics, old folk or babes in arms. The new science of physics, which established that gravity determined motion and stability in the universe; and the simple question, "*Unde malum?*" – where does evil come from? Together, these two challenges brought down the belief in divine Providence. It was abandoned even by theology.

Whoever believes in *providentia* needs no concept of sustainability. For the future lies in God's hands. But when this faith begins to falter, an abyss opens up. "No saviour from on high delivers." What can take its place? Are God and Nature the same thing? Is evolution a kind of Providence of nature? Can human beings take over the *gubernatio*, the steering of nature? Will human reason find in the laws of nature the keys to the mastery of the Earth? When the *prima causa* (God) disappears, how do the *secundae causae* operate? How does the *cooperatio* between humans and nature function? Harmoniously? Or do human beings become the guardians of foresight and precaution? If so, what kind of guardians? Like the head of the house, the 'husband', the faithful steward who administers the entire household for the benefit of all, including the servants, the beasts in the stable and the plants in the fields, and passes it on to the following generation enhanced? Should this role be taken over by a strong, absolutist state? The term for the 'welfare state' in modern-day France is still *État Providence*. Or does the 'invisible hand' of the market regulate everything? In the great melting-pot that was the early Enlightenment, all of these ideas were still swimming around together until discrete bodies of opinion finally coalesced and solidified.

And today? Suddenly the vocabulary of the doctrine of *providentia* is back – in the global discourse of sustainability. The Latin word *conservatio* has been taken over into English and French almost unchanged. In the sense of 'conserving utilisation' it has become the opposing term to depletion and environmental degradation. An international network of environmentalists and ecologists, the International Union for Conservation of Nature (IUCN), decided in 1980 that 'conservation' and 'sustainable development' were the same thing. International policy think tanks fiercely debate issues of global governance or earth systems governance – the old idea of *gubernatio* in new guise. All over the world, conferences are debating a new integration of ecological and social impulses. That used to be called *concursus*. And what do we understand by *terminatio* – the limitation of evil? Today's arguments are about limits on toxic chemicals, about limitation of per capita CO_2 emissions, and – once again – about the limits to growth. Finally, films and books create an endless stream of new images of annihilation, of the destruction of the planet. The existential fear of the apocalypse, which was the undertone accompanying the belief in Providence, has not been exorcised.

Providentia – does it have any currency today? There can be no doubt that the secularisation of the idea is irreversible. Rather than believing that "He's got the whole world in His hands" we are more likely to believe in the message "the world is in your hands" from the advertising of the global credit-card companies. However, the question of whether we will succeed in saving the planet is by no means decided. Shortly after the terrorist attacks of 9/11, the German philosopher Jürgen Habermas recommended that secular society should try to maintain a "feeling for the expressive power of religious language" where it is "a bearer of semantic content . . . that eludes the explanatory force of philosophical language". In the command to "work . . . and take care of" the Earth (*Genesis* 2:15, New International Version), the biblical Garden of Eden narrative gives us a simple formula for sustainability. In the Hebrew version the words used here are *abad* (to tend, to cultivate) and *shamar* (to exercise care over, to protect, to maintain). Which brings us very close to the formulation: sustain and develop – sustainable development.

The murder of Mother Earth

> Pale of face she appeared in the witness stand. She was dressed in green. From her eyes flowed tears. There were injuries to her head, her dress hung in tatters, and one could see that her body had been pierced through many times . . . covered in wounds and blood . . . No longer any trace of grace or beauty.[5]

We are talking again about *mater terra*. But this time it is about rape and murder, this time she is the victim. A short story, written in the Latin of the Humanists, which transports the reader into the era of the Renaissance – and into a boom period for early European capitalism. In 1477, half a century before the Conquistador Pizarro landed on the coast of Peru and set off with 180 men in search of the Inca land of gold, a European 'Eldorado' was conjured up in the silver-rich Ore mountains ('Erzgebirge') which straddle Saxony and Bohemia. In nearby Chemnitz, a teacher and scholar by the name of Paulus Niavis (Paul Schneevogel), sickened by the silver fever around him, put together an allegorical story. It tells of a court case. The charge: the rape of Mother Earth. The little book, which appeared in 1492, bore the title "Iudicium Iovis – the court of Jupiter, held in the Valley of Beauty . . . "

In the Dresden Museum of Mineralogy and Geology one can still see relics of the time Paulus Niavis wrote about: two deeply-scored nuggets of silver. One is the size of a fist, 600 grams of pure silver. The other weighs nearly seven kilograms and is composed mainly of silver-glance (argentite). These samples of different ore grades come from the time of the great silver rush, the spectacular find in the year 1477 in the western Ore mountains. According to tradition these are fragments of the legendary table of solid silver at which Duke Albrecht and his followers are supposed to have dined "in the lap of the earth", in a shaft of the St George mine at Schneeberg ('snow mountain'). But perhaps they are simply presents to the sovereign from the mine operators on his return from a pilgrimage to the Holy Land. The sudden wealth had its drawbacks. Silver fever drew thousands of prospectors "up to the snowy mountain". Intoxicated with greed, they drive shafts and tunnels into the flanks of mountains, clear the woods, divert streams and poison them with the mine run-offs, polluted with lead and arsenic. The smoke wreathing from countless charcoal kilns and smelters fouls the air. The brief silver fever leaves deep scars on the face of the landscape.

In the early summer of 2009 I hiked to the 'sites of memory'* of that episode. The hilltop in the southern part of Schneeberg offers a wonderful panoramic view over the town. This baroque construction lies on a hill on a broad valley floor in the northern foothills of the Ore mountains. At the spot where the tower of the huge Late Gothic hall church rises up, the first shafts and tunnels were driven into the mountain in 1477. Today this history is a source of pride. A guided 'mining walk' connects the relics: shafts laid out with beautiful brickwork, lovingly restored headhouses, a stamp mill for processing the ore, a pond which collected the water which in turn drove the mining machinery.

Over the next few days, my route took me along the river Mulde to the ridge of the Ore mountains, then from Johanngeorgenstadt further on on the Czech side over the high plateau to Jáchymov, formerly known as St Joachimsthal. Three days' hiking in a wonderful low mountain landscape: quiet mountain forests, panoramic hills, trout streams, moorland, bizarre rock formations. On the

---

* 'Sites of memory' (*lieux de mémoire*) is a term established by the French historian Pierre Nora during the 1980s. It signifies the relationship between the sites of historical events, developments etc. and the collective memory and identity of a nation, a society, or a group.

naked rock of the steep hillsides, here and there the contact zones between the granite and the greywacke rock come to the surface. These are the points where, deep down in the lap of the earth, the veins of silver and other metals were formed. On this terrain one repeatedly stumbles upon traces of ancient mining activity: the silently flowing water of a ditch, the bricked-up mouth of a tunnel, spoil piles and old mineshafts. Some of them go back to those boom years around 1500. Everything is overgrown, covered with luxuriant vegetation.

Shortly before Jáchymov my path leads over the 1,000-metre-high basalt hill-top of the Plešívec. From up there, the view curves along the entire crest of the ridge as far as the twin peaks of Fichtelberg and Keilberg, then down into the narrow wooded valleys of Jáchymov and Hroznětín (Lichtenstadt) and south-wards onto the plain of the Eger Graben. Somewhere around here, at the foot of the Plešívec, was where Paulus Niavis set "the Valley of Beauty", the scene of the court case of which he tells in 1492.

The plot is as follows. A hermit from Lichtenstadt happens upon a trial being held by the ancient gods near his cave. A tribunal, chaired by Jupiter, sits in judgement over *homo montanus*, the miner. The charge: desecration of Mother Earth by penetrating her bowels – the womb, the matrix. Her advocates, Mercury and Minerva, argue that the Earth brings forth fruit every year with which she nourishes and sustains (*alit atque sustentat*) all living things . . . "But not satisfied with this kindness, man penetrates into his mother's bowels, ransacks her body, wounds and damages all the inner parts. In the end he tears the entire body apart and completely paralyses its powers." Can the Earth withstand such treatment for long and endure in the face of this "frenzy of humankind"? Turning to face the accused, Mercury shouts at him:

> You murderer! Look at her! She who not only feeds you and keeps you alive (*nutrit et in vita conservat*), but who even takes you back, after you die, into her womb, from which you came . . . Is there no trace of love in you for she who bore you?

The miner, the accused, shows no sign of guilt. His defence is calm and rational and founded on economics. Resources are distributed unequally among the regions of the world. In order to trade them you need money, e.g. silver. This metal, given to Man by the gods, has been stored up by the Earth in her womb and thus wilfully denied to him. So his forcible invasive action is justified. Not everything which goes beyond "the essential needs" is wicked luxury. Silver can also be used to help the poor and to maintain order. Without the work of the miners there could be "no state and no sociable coexistence". Trade between peoples would cease. "We would live again in the woods, like animals." He doesn't acknowledge the condition his victim is in. The accusation that he is destroying the foundations of his own existence he simply ignores.

In this story from the Ore Mountains, as in the Canticle of the Sun two centuries earlier, nature appears as life-giving Mother Earth. But now she must be protected against the actions of humankind. For through his 'shock strategy' (to use a contemporary term) he is destroying the foundations of his own existence. And in this text from the early modern period, too, the vocabulary of sustainability appears, encapsulated in key terms from the doctrine of *providentia*: *sustentare* and *conservare*. The Earth's advocates use it. The accused cleverly adopts this rhetoric in his own defence: he claims the right to act as he does in order to fulfil the divine command to 'work . . . and take care of' the Earth.

And it brings him success. The miner gets away with his crime. It is however only a partial acquittal. In the reasoning given for the verdict, the gods declare that it is the "fate of mankind . . . to ransack the Earth". But the verdict contains a stark warning. If the miner carries on with his unsustainable practices, the Earth will sooner or later take her revenge and destroy him. The bodies of the invaders will be swallowed up, suffocated by evil gases, subject to all kinds of dangers. This sombre prediction brings to a close this story from 1492 – the year Columbus landed in the New World and the rape of *mater terra* reached a new level of brutality.

A surprising discovery: the pre-Columbian culture of the Andes has a counterpart to the little story from the Ore Mountains. Eduardo Galeano, the chronicler of South America, documents it in his epic book *Memory of Fire*. The episode takes place at the 'Cerro Rico', the 'rich mountain' at Potosí in today's Bolivia. "Before the conquest, in the day of the Inca Huaina Capac, when the flint pick bit into the mountain's veins of silver a frightful roar shook the world. Then the voice of the mountain said to the Indians: This wealth has other owners." The Incas never touched the unprecedentedly rich silver deposits at this mountain. The image of *mater terra* had an equivalent in their culture: Pachamama is the earth goddess, who rules the world together with Tuta Inti, the Sun god, and gives all creatures life and nourishment. It was the Conquistadors who commenced the exploitation of the mines at Potosí. Europe was flooded with silver bars. The mining industry in the Ore Mountains went into a prolonged crisis. But the slopes of the Cerro Rico, Galeano wrote, soon ran red with human blood. The victims of this merciless exploitation ran into the millions.

In April 2009, the Bolivian President Evo Morales appeared before a full meeting of the United Nations General Assembly in New York. For the indigenous movement, he said, harmony with Pachamama – Mother Earth – is sacred. "Mother Earth gives life. . . If we talk, work and fight for the wellbeing of our people we first have to fight for the wellbeing of Mother Earth." Was this a return to his own cultural roots, or the tactic of a populist? When it comes to oil exploration, the Bolivian state-owned consortium Petro Andina is as ruthless in its dealings on Indian territory as its predecessors were.

Our story from the Ore Mountains also has a contemporary legacy. The landscape is littered with scarcely-healed wounds. Hiking over the mountain ridge takes one past fenced-in, conical heaps thinly covered by young trees, and hillocks of cleared reddish scree. In the tourist mine at Johanngeorgenstadt, the guide takes a blackish shiny stone from its glass case and elicits with it a ticking noise from a Geiger counter: uranium ore. Until the Eastern Bloc collapsed, the region either side of the ridge of the Ore Mountains was one of the biggest sources of uranium in the world.

The history of this mysterious element is linked with the Ore Mountains from the very beginning. Traces of uranium have been found, using modern research methods, in ore samples from the silver rush of 1477. Three centuries later, in 1792, Martin Klaproth, a chemist from Berlin, succeeded in isolating a hitherto unknown element from samples of pitchblende (uraninite) taken from Johanngeorgenstadt and St Joachimsthal, and called it uranium. Another hundred years later, the Polish researcher Marie Curie demonstrated the existence of rays which she called radioactivity emanating from material from Jáchymov. The first Soviet atomic bomb in 1949 was detonated using uranium from Johanngeorgenstadt. In the early years of the Cold War, the Soviet secret service established a hermetically-sealed zone for the extraction of the uranium deposits in the valleys and hilltops of the Ore Mountains. The 'nuclear winter', that nightmare vision of the early 1980s, would have been set off with the help of radioactive material from the Ore Mountains. Luckily, it didn't happen. Otherwise, the sombre warning given by Mercury and Minerva, the advocates for *mater terra*, would have come about – representing a final judgement.

A historical irony. The *Joachimsthaler Guldengroschen*, or "golden penny from Joachimsthal", made out of the high-grade Ore Mountain silver of the first boom period, soon had its name abbreviated to 'Thaler'. In turn, in 1776, the US dollar, later to become the currency of globalisation, was named after this coin. Today, the solid Renaissance structure of the Jáchymov mint holds a museum. At its entrance is a coin-punch. With one blow of the hammer, visitors can stamp their own Joachimsthaler. They are made, however, of cheap aluminium.

CHAPTER FIVE

A European dream

These are tumultuous times. Much of the world is going dark, leaving many
human beings without clear direction. The European Dream is a beacon of light in
a troubled world. It beckons us to a new age of inclusivity, diversity, quality of life,
deep play, sustainability, universal human rights, the rights of nature, and peace
on earth. We Americans used to say that the American Dream is worth dying for.
The new European Dream is worth living for.[1]

These are the closing sentences of a book published in 2004 by the US sociologist
Jeremy Rifkin on *The European Dream*. Its message – at the same time comforting
and challenging – raises a few questions for Europeans. Is there really such a
thing as the European Dream? If there is, how does it differ from the "American
Dream"? Could the answer lie in the theme of sustainability? Archaeological
research into that question has to dig very deep. In the old Europe, the overthrow
of the established cosmology and associated beliefs began with the (literally)
revolutionary insight that the world revolves.

The astronautical perspective 1440-1634

Seen from far away, the earth appears as *lucida stella* – a bright star.[2] As animal –
that is, as a living being. Its form is noble and spherical. It is the unique *habitatio*
– modern ecologists say 'habitat' – of human beings, animals and plants. Nicholas of Cusa (Cusanus) was convinced of this already. A theologian and natural
philosopher from the Mosel region of Germany, he calmly describes in 1440 his
cosmic vision. The universe is potentially infinite. It has neither a boundary nor
a solid body as a central point. "Even if there are inhabitants of another kind on
other stars . . . with regards to the intellectual natures a nobler and more perfect
nature cannot, it seems, be given than the intellectual nature which dwells both

here on Earth and in its own region." Here, the perspective from above, the astronaut's perspective, has been anticipated by the imagination, 500 years before the first space flights. The verdict is the same: the Earth is the most beautiful star in the heavens.

For Cusanus there was no doubt that divine Providence was "necessary" and "unchangeable". However, the fascinating new conceptual models of the universe which Nicholas of Cusa and his contemporaries constructed triggered an existential unease. If the Earth is not the fixed hub of the cosmos, but a floating speck of dust at the edge of space, does it not follow that the universe is far too big, too old, too multiform to be there for Man's sake? What if Man has lost God's grace? If divine Providence is no longer operating, how can Man endure in this immeasurable, overpowering, indifferent cosmos? And more: is there any possibility of positive development?

Development, for Nicholas of Cusa, is first of all a geometrical term: *Linea est puncti evolutio.* The line is the development – literally, the unrolling – of the point. As a synonym for *evolutio,* Cusanus uses the word *explicatio,* derived from *plicare* – to fold. In this era, development is the unfolding of the immanent qualities of things. Neither more nor less. Its symbol is the seedcorn from which the young plant grows. Development is highly sensuous. The artists of the Renaissance were fascinated by how clothing falls into folds on the human body. Painting its forms, achieving the plasticity of the geological folds of the high mountains while retaining the lightness of the airy formations of a cloud in the sky, was the test of a true master. It is worthwhile trying to recapture these subtle images and expressions today. It is an exercise which may lead us to a deeper grasp of the combination term 'sustainable development'.

Johannes Kepler, too, had a dream. Two hundred years after Nicholas of Cusa, towards the end of the Renaissance, this astronomer (who had studied Cusanus's book at the university of Tübingen) tells the story of an imaginary journey to the Moon. His *Somnium* (Dream) appears posthumously in 1634.[3] Kepler, on whose calculations of the orbits of the planets NASA's moon flights were still based, clothes his description of the heliocentric model of the universe in a dream narrative, in which fantasy is mixed with what were then revolutionary scientific insights.

This is how the story goes. After many years of study under the Danish astronomer Tycho Brahe, the narrator has returned to his home, which in the dream is Iceland. As he tells his mother about the work in Brahe's observatory, she – a herbalist who had cast him out of the family home many years earlier – interrupts him to mention her friends, the "wise spirits", who have withdrawn to cold, dark Iceland and who set out from there on long journeys – even to the Moon. Curious, the son asks his mother to admit him to the circle. One spring night, she takes him with her to a crossroads where with appropriate ceremonial she conjures up the spirit. Hardly have the two of them covered their heads when a hoarse, unearthly voice begins whispering. The voice recounts, in Icelandic, the story of a journey to an island "50,000 German miles up in the ether . . . the island of Levania" – the Moon.

A good 350 years before the beginning of space flight, Kepler now unerringly identifies the trickiest problems it poses: the massive thrust required for launch, the transition to weightlessness, the extreme cold in space, the impact when landing on the Moon. For a journey like this, you don't need somebody "who is lethargic, fat or tender . . . No men from Germany are acceptable. We do not spurn the firm bodies of Spaniards. We especially like dried-up old women, experienced from an early age in riding he-goats at night or forked sticks or threadbare cloaks, and traversing immense expanses of the Earth." There follows a description of the surface of the Moon and of the weather conditions prevailing there. Then Kepler looks backwards. For the "greatest of all spectacles" available to the inhabitants of the Moon is the view of "Volva", the Earth. Although she appears not to move at all, in fact she "turns like a wheel in its own place and displays a remarkable variety of spots, one after the other moving along constantly from east to west." From the Moon, one can discern clearly on the globe, slowly revolving on its axis, the lighter half, namely the oceans, and the darker, patchier continents. The land mass of Africa appears on the rotating planet like an image of "a human head cut off at the shoulder". Europe, "a young girl with a long dress", bends down towards this head "as if to kiss it", at the same time beckoning towards her by means of her outstretched arm "a leaping cat" – namely Britain.

If one now looks beyond Earth into space, one sees the stars moving and becomes aware of the complex system of their motions. Like the Earth, they occupy orbital paths around the Sun which in the infinity of space can be mathematically calculated and geometrically represented.

The first-person narrator of Kepler's *Dream* is an enlightened European of the 17th century, like the author. Tycho Brahe, the Danish astronomer, was his teacher in Prague. Kepler's mother had been persecuted as a witch in Weil der Stadt in Württemberg. Only his forceful intervention had saved her from the stake. One year after her release from prison, she died from the effects of the torture. The didactic purpose of the story is clear: to make the new heliocentric model of the universe comprehensible through the use of different perspectives and benchmarks.

Kepler sees himself as a "Priest of God at the Book of Nature". Through investigating the laws of planetary motion in the solar system, he hopes to discover the structure and rules of celestial harmony. The belief in *providentia* still holds. But it is on the verge of a crisis. "The eternal silence of these infinite spaces terrifies me," writes Kepler's contemporary, the French mathematician Blaise Pascal. The effect of the shock oscillates between fascination and bewilderment, humility and hubris.

Out of joint

"The time is out of joint," cries Shakespeare's Hamlet from the stage of The Globe theatre in London. A pirated print copy of the play appears just after 1600. This is the same time at which Kepler, casting horoscopes for the imperial court in Prague, discovers a "new star" in the Milky Way – a supernova.

What is it that is "out of joint" at this precise time? If the new view of the cosmos now emerging was disconcerting, then the experience of chaos here on earth was traumatic. The bloody power games of the rulers, endless campaigns, wars of civil and religious wars are laying waste to the old Europe. In the wake of the wars come the Apocalyptic horsemen: the Black Death, the plague, then famine. Entire regions are pillaged and depopulated. Mass graves everywhere. The terror of the Inquisition, the massacres of women under the pretext of witch hunts, the insane butchery perpetrated by the Conquistadors in the Indian empires of America – to many people it is clear that an order and a value system are collapsing before their eyes. Several generations of Europeans stared into the abyss.

Even the climate seems to be out of joint. Historians speak of the 'little Ice Age' which gripped Europe in the 17th century. A temporary decrease in sunspot activity at this time resulted in lower solar-radiation intensity. Central European chronicles tell of grim, cold winters with heavy snow followed by widespread ice, flooding and late frosts. Summers are either cold and wet or extremely hot. Hailstorms are frequent, poor harvests almost the norm. Landscape painters depict the village ponds of Flanders, Brabant and Scotland teeming with ice-skaters. Even the Venice lagoon is reputed to have frozen over in some winters. The opening of 'Winter' in Vivaldi's concerto *The Four Seasons*, depicts this in sound in 1723.

"The time is out of joint," says Hamlet to his two companions, with whom he had studied at the university in Wittenberg. Bewildered and racked with doubt, he curses his fate in having been "born to set it right".

The great novelty of the era, wrote the German philosopher Hans Blumenberg, was the compulsion to "human self-preservation". In the age of the Enlightenment, self-conservation – *conservatio sui* – became the main concern of European philosophy. The concept of *conservatio* moves out of the theological doctrine of *providentia* and into the heart of philosophical and economic discourse. The groundbreaking philosophical systems of the 17th century contain two distinct models, in embryonic form, of conservation – and thus paved our way to the modern concept of sustainability.

The Descartes Model

"Cogito, ergo sum." I think, therefore I am. René Descartes based the hope of a new beginning on this philosophical 'ground zero'– the certainty of being a thinking subject. All power to Reason! – this seemed like the only hope in a time when the belief in a divine Providence was crumbling. He formulated this principle in the *Discours de la Méthode*, which appeared in Leiden in Holland in 1637. But the insight had first come to him in a dream almost two decades earlier.

Even the great rationalist dreamed. On the eve of St Martin in 1619, while stationed for the winter in Neuburg on the Danube as a 21-year-old volunteer in the service of the Duke of Bavaria, he has a series of dreams. The first is a nightmare. Tormented by evil spirits, he suffers a paralysis of the whole of the right side of his body while walking along a road. He limps onwards on his left leg, is spun around three, four times by a powerful gust of wind. He drags himself forward, no longer sure of the ground beneath his feet, terrified of collapsing at any step. Then he sees in front of him a school building and decides to take refuge there. In the final dream, he opens a book of poems and reads this verse by the Roman poet Ausonius: "Quod vitae sectabor iter . . . ". Which path should I follow in life? Descartes carried about with him for the rest of his life the notes he made about his dreams that night. His notebook has not been preserved, but Leibniz was able to read the original.

Here, everything centres on the absence of solid ground in our existence and the search for a new foothold. "In the dream," the media theorist Friedrich Kittler says of the storm-motif, "the subject becomes an unextendable point or better, midpoint, around which one's own body, as a three-dimensional *res externa*, describes the geometric figure of a circle. All of Cartesian philosophy deals with this *res cogitans* and that *res externa*." [4] In Descartes' own words: ". . . there are no points really immovable in the universe . . . nothing has a permanent place unless in so far as it is fixed by our thought."

"Je pense, donc je suis." He selects this initial, indubitable and irrefutable certainty as the starting-point of his *Discours de la Méthode*, the search for a method of establishing truth. "I think, therefore I am" was an unprecedented blow struck in the cause of freedom. The autonomous subject acts independently of the authority of the Church or of an absolute ruler. But more than that: this very thought liberates itself from the bonds which tied it to a body, indeed to any part of nature. The disembodied *res cogitans* and the material world are radically divided from each other. Nature, even the human body, is *res extensa*, extended matter, a mechanism. Made by a watchmaker-god who, having wound up the mechanism, disappears from the picture. The now liberated thinking subject focuses on the objective of self-preservation. The method: gaining mastery over nature, taking possession of it, classifying it and making it useful. The means by which this is achieved is rational thought: only acknowledging as true what is evident and demonstrable. Breaking something down into as many constituent parts as necessary, dissecting, analysing, measuring, rearranging and reconstructing – this is the Path of Enlightenment. By these means Descartes hopes to find "knowledge that would be very useful in life". His aim is ". . . to find a practical philosophy, by means of which, knowing the force and the actions of fire, water, air, the stars, the heavens, and all the other bodies that surround us, just as distinctly as we know the various skills of our craftsmen, we might be able, in the same way, to use them for all the purposes for which they are appropriate, and thus render ourselves as it were, masters and possessors of nature (*maîtres et possesseurs de la nature*)." [5]

Descartes' path leads to the idea that man can and must control nature (he prefers the word 'matter'), align it with his purposes, rearrange and reconstruct it. Nature in this view is no more than a resource repository, to be arranged and utilised according to rational criteria. The consequence was a radical devaluation of nature. The separation of mind and matter paves the way for what the American historian Carolyn Merchant called the "death of nature". Whether that was the intention is open to question.

Something of the 'bon père', the good housekeeper who manages responsibly what is entrusted to him, can be heard still in Descartes' 'maître'. The "possesseur" is not "propriétaire", untrammelled proprietor, but in temporary possession, a leaseholder who has to pass on, preferably intact and undiminished, what he temporarily possesses. Descartes specifies the overarching aim of his philosophy as "the conservation of health" (*conservation de la santé*). In the language of the time, this means more than the prevention of illness. *Santé* refers to a state of harmony between the body, the soul and external nature, including an intact environment. Seen in this way, man as 'maître et possesseur' must also be *conservateur* – preserver – of nature. But is control and possession compatible with conservation in the long run? This combination, as we shall see, holds great potential risks.

The Spinoza Model

In the autumn of 1649, Descartes leaves Holland after a stay of twenty years and travels to Stockholm at the invitation of Queen Christina of Sweden. He doesn't survive the harsh northern winter: in February 1650 he dies of pneumonia. The following year, a young man from the Amsterdam Jewish community named Baruch Spinoza begins reading Descartes' works. Ten years later, accused of 'materialism', Spinoza has been issued by the Jewish authorities with a writ of *cherem* or *banvloek*, a form of excommunication. He is grinding lenses to earn a living, lodging with a doctor in the village of Rijnsburg near Leiden. He begins his magnum opus. It is in one sense fortunate that this work, *Ethica*, first appears posthumously in 1677.[6] The Dutch rabbis who excommunicated him, as well as the Calvinist Church Council of Amsterdam and the 'stadholder' William of Orange, all regarded him as an atheist. The inscription on Spinoza's signet ring reads 'Caute' – be cautious.

A highly sophisticated theory of sustainability can be distilled from Spinoza's *Ethics*. *Suum esse conservare* – to preserve one's own being, self-preservation – is the fundamental human drive (*conatus*). This is the hypothesis on which the work is based. From the 'great Cartesius' Spinoza has adopted the maxim of "deducing only from principles which require no external proof". He also shares his teacher's aspiration to survey and map human thought anew, in geometrical order . But Spinoza posits a radical alternative to Descartes. He accomplishes the greatest possible upward revaluation of nature: he declares that God and nature are one and the same.

Deus sive natura ('God or nature') is the living Ur-Being, which exists in itself, and of itself and of necessity fulfils the totality of its potential. Nature incorporates attributes of God, as the mirror image, as it were, of this invisible Being. All individual phenomena are modes of these attributes. For Spinoza there is only one universal substance. All lesser 'entities', including human beings, are modes or modifications of this single universal substance. This substance cannot be known because it "stands beneath" matter, but it is in everlasting pulsing creation. We are but reflexes of this creativity, and we react from necessity, as determined by our nature. It is only from these affects, or feelings, that we know and understand ourselves. The Cartesian dualism of mind and body, of thought and matter, is resolved: the thinking substance and the extensive substance are one and the same.

God = Nature. The two terms of the equation can of course be reversed: Nature is God. Here is the first cause of all existence, including thought. The more we know of the individual phenomena of nature, the more we know of God. Spinoza now considers nature from two perspectives. *Natura naturata* is 'caused', created, that is to say empirical nature. He distinguishes from this *natura naturans*, the living, active and productive force which is the causal principle within *natura naturata*. The distinction is essential. Nature as *natura naturata* is subject to the human will. It can be manipulated and reproduced. But the vital powers of *natura naturans* are all-powerful and absolutely unavailable, indisposible and intangible. They are the fullness of life, the force of life itself.

Spinoza categorically rejects the doctrine of Providence. The idea that 'God made all things for man' and that there is 'a ruler of the universe . . . who has arranged and adapted everything for human use' is a prejudice. Rather, he believes that "everything in nature proceeds from a sort of necessity and with utmost perfection". Substance cannot be created, but exists, he writes in a letter of 1661 to Henry Oldenburg, Secretary of the Royal Society in London. From this he deduces that "all substance must be infinite or supremely perfect after its kind". With this, man's claim to mastery collapses. The classification of natural phenomena into good and evil, useful and harmful, simple and sophisticated is invalidated. In its place comes the indivisible web of life. Against Descartes' enthronement of man as the master and owner of nature, Spinoza insists that man – the human body and the human mind – is also a part of nature. And he goes on to say that each part of nature agrees with its whole and is associated with the remaining parts. He is therefore by no means abandoning the project of human self-preservation, but embedding it in the wider, ecological context.

Suum esse conservare, the preservation of one's own being: this natural drive is the starting-point of every desire, and thus also of economic activity. Since the expulsion from Paradise, human beings are responsible for their own welfare. Spinoza sees the expulsion not as a punishment, but rather as a step on the path to becoming fully human. As there is no further divine intervention, and no continuous divine providence, human beings have to provide for themselves. This is in fact an important element of their freedom and dignity. Ensuring economic security, however, can only be achieved in cooperation with nature. The treasures of nature are provided for us, we do not create them. Our freedom consists in bringing the endeavour of the better part of ourselves in harmony with the order of nature as a whole. Where we are successful in this, we will assuredly acquiesce in what befalls us and in such acquiescence will endeavour to persist.

Spinoza brings into play infinite duration (*duratio*), the existential dimension of time. Each thing, as far as it can by its own power, strives to persevere in its being. This includes a drive for development. For example, it is rational to prefer a greater future benefit to a smaller present one. Responsibility for the future thus becomes a constant of our thinking. It is in the nature of reason to perceive things under the aspect of eternity (*sub aeternitatis specie*).

What does this mean for the structuring of a stable and enduring, a sustainable society? Reason requires us to see self-preservation as linked not just to the preservation of the essential natural foundations for life but also to the welfare of others. It is apparent that people can provide for their wants much more easily by mutual help and that only by uniting their forces can they escape from the dangers that on every side beset them. This is Spinoza's riposte to the English philosopher Thomas Hobbes, who had taught that in the state of nature, Man is a wolf to Man. "Man is to Man a God", says Spinoza. In opposition to the lupine laws of free competition he declares that "in the state of nature, no one is by common consent master of anything, nor is there anything in nature, which can be said to belong to one man rather than another: all things are common to all." Spinoza believes in the just distribution of goods and in the *potentia multitudinis*, the democratic power of the many. "[T]o Man there is nothing more useful than Man, and nothing more excellent for preserving their being can be wished for by men, than that all should so in all points agree that the minds and bodies of all should form, as it were, one single mind and one single body, and that all should with one consent, as far as they are able, endeavour to preserve their being, and all with one consent seek what is useful to them all." At this point in the *Ethics* he formulates the basic idea of social justice: to "desire for themselves nothing, which they do not also desire for the rest of mankind". He derives from this a complex conception of the quality of life. In his language this means: to be happy, to act properly, to live a good life. Desire, sadness and joy are the three emotions (affects) which move human beings. The core of happiness or well-being (*beatitudo*) is joy. Joy – or love – is the affect which decisively increases the capacity of mind and body, their power to act, their self-sufficiency.

In Descartes' philosophy, the individual rational human being ascended to the position of master and owner – thereby also keeper, guardian – of nature. Spinoza celebrates the multitude, the free association of people, who develop in harmony with nature, including their own nature. The conception of sustainability which evolved in the early Enlightenment oscillated between these two poles. It has continued to do so ever since.

The best of all possible worlds

'Another world is possible.' The slogan used by today's critics of globalisation would have elicited from the great thinkers of the early Enlightenment – according to temperament – either a smile or an angry frown. The philosopher Gottfried Wilhelm Leibniz, born in Leipzig in 1646, described the universe mapped out by the astronomers of the Renaissance as "the best of all possible worlds". Known as 'The German Plato', he was a member of the Académie Française, a correspondent and rival of Newton, an early student, interlocutor and critic of Spinoza. His formulation is easily misunderstood. Is it a justification of the prevailing state of affairs? Does it attempt to claim that the existing social order is untouchable, or even sacred? Does it thereby contradict 'another world is possible'?

First of all, 'world' for Leibniz is the entire comprehensible and perceptible world, including natural phenomena and natural laws. In contemplating this whole, he queries why, at the moment of Creation, out of the infinite multiplicity of possible worlds, God chose to launch precisely this one. His answer is that God acted rationally. His creation is "at the same time the simplest in terms of principles and the richest in terms of phenomena" – and in terms of potential. This world offers the greatest range of possibilities for action with the minimum of regulation. It is the best possible precisely because it allows the freedom to develop the world's potential and to perfect its qualities. The French philosopher

Gilles Deleuze interprets Leibniz's formulation to mean that it allows human beings "a subjective production of novelty – in other words, creativity". "The best of all possible worlds" is not a description of a situation, but rather an incitement to be creative. It embodies the assurance that the goal of perfectibility is built into the objective world as it exists, and therefore attainable. And that would not be so very far from 'another world is possible'.

To maintain or conserve this best of all possible worlds – and perhaps to develop it further? Does that not sound familiar? Leibniz comes very close to the coupling of conservation with development which underlies sustainable development. And indeed he contributed substantially to establishing the terms *evolutio* and development – or rather *développement*, as he used the French term – in the vocabulary of the European early Enlightenment.

The Latin word *evolutio* originally referred to the unrolling of a scroll or manuscript. Within the French word *développer* lies the Latin *develare* – to expose or unveil. The English form of the word – evolution – appears for the first time in 1670 in the *Philosophical Transactions*, a journal of the Royal Society – and already then in a biological context. Leibniz entered into this debate. Strictly speaking, he said, there is no such thing as procreation, nor death, but only the extension and contraction of already extant forms of life. All things are pre-formed. Only that which was previously *enveloped* can *develop*. Development, or evolution, is nothing other than the gradual appearance of the fundamental and primal attributes of things, brought about not only by the operation of innate driving forces but also through external influences. Development cannot be continued indefinitely. Growth and death belong together.

In all of this, Leibniz takes for granted the fundamental benevolence of Creation; or as he puts it, the "pre-established harmony". The world is not the "vale of tears" of the medieval Scholastics. To be sure, human beings are no longer the centre of the universe. But instead of worrying obsessively about the evil we suffer, we would do better to recognise this harmony in the universe, to acknowledge it and to use this knowledge for what might be called the *sustainable development* and the perfection of our world.

Leibniz's guiding principle is, as it was for Spinoza, the happiness of the human race. The universal aspiration to happiness in this life – as opposed to salvation in the next – is a new agenda introduced by the early Enlightenment. In 1776 it enters into the American Declaration of Independence, and into the 'American Dream', as the "pursuit of happiness". However, from the very beginning it is coupled very closely there with images of unbounded material wealth – or to put it another way, with the dollar.

Leibniz was an inventor. Just as Kepler laid the foundations for the calculation of satellite orbits, so Leibniz devised the binary system which enables us today to navigate the virtual worlds of cyberspace. His principal interest, however, lay in solving acute practical problems, and these concerned above all the question of energy supply. Leibniz constructed a machine designed to pump water by means of wind power out of the mines of the Upper Harz mountains which were the principal source of income for his Hanoverian sovereign. It didn't work. But this failure only strengthened his interest in the experiments of his Huguenot friend Denis Papin, who around 1690, while in the service of the Landgrave of Hessen in Marburg, constructed a simple model of the steam engine. In an essay which appeared in the scientific journal *Acta Eruditorum* published in Leipzig in 1690, Papin (incidentally, also a member of the Royal Society in London) described this "new method of producing significant power at low cost".

His piston steam engine was still only a toy. But the challenge of finding new means of supplying energy was being posed with ever greater urgency. For Leibniz and his generation, this was the key to the future. The nightmare which haunted the whole of Europe at that time was a shortage of the central resource, of wood. The efforts towards the self-preservation of society, involving a huge expenditure of mental energy, were focused on the *conservatio*, the conservation, of the forests. The idea of sustainability was conceived in the networks of the early Enlightenment in Europe around this key question. The terminology, however, had already begun to emerge in Renaissance Venice.

CHAPTER SIX

Virtuosi

The forests of la Serenissima

Seen from the water, Venice appears to be made entirely of marble. At least, that's how it seems to anyone taking a vaporetto on Route No 1 down the Grand Canal. The Rialto bridge, the palaces to left and right, everything is made of the hard, white or red, mildew-spotted stone. The magical light of the city is due to the mirroring reflective action of water and marble. But the city on the lagoon stands on foundations made of wood. The fact that the stone Palazzi are supported by rotting wooden piles remains hidden from the eyes of visitors. They see only the light-brown pile-ends which protrude from the water at the moorings. Beyond the last big bend of the canal a wooden bridge comes into view, the Academy bridge. The canal opens out to the lagoon. Next stop: 'Salute'. A magnificent flight of steps leads up to the baroque edifice of Santa Maria della Salute, crowned by two monumental domes, its walls clad entirely in brilliant white marble. It will have a role to play in our Venetian sustainability story. The vaporetto now makes its way alongside the long flat building of the Customs House and towards the heart of the city. One after another, the slim Campanile, the monumental curves and arches of San Marco, the pale marble south façade of the Doge's Palace, once the centre of Venetian power, for a brief moment the Bridge of Sighs, symbol of the dark, cruel side of that power, all come into view.

Two more stops, and we reach the real centre of power of la Serenissima: the Arsenal. A short walk from the mooring brings us to the entrance. To the left, a Renaissance building blocks the landward access. Beyond the triumphal arch with the winged lion above the portal, a broad open site stretches out, imposing and sinister still today. In its prime, thousands of 'Arsenalotti' worked here day and night, on the largest industrial complex of the old Europe. Roofed floating docks, a ropeworks, compass workshop, rudder workshop, the pitch distillery which already by around the year 1300 had conjured up for Dante a vision of hell, of the 'Inferno', of hell – the Arsenal was once the famous shipyards of la Serenissima, the backbone of Venice's sea power.

Wood was *materia prima*, the principal raw material. The *tese*, the timber warehouse of the shipyards, was, at 140 metres long, one of the biggest buildings on the site. Demand was enormous: high quality oak for the hulls, mature straight-grown oak and firs for the masts, beech for the oars of the galleys, softwood for the distillation of pitch – these were the priorities at the Arsenal. Only then did the needs of the city's residents come into consideration: oak and larch trunks for the piles and breakwaters, charcoal for the glass furnaces at Murano, firewood for the households, for cooking and for heating in the icy Venetian winters.

Until the late Middle Ages there seems to have been a functioning free market in wood. But in 1458, the Senate set up a new office, the *provveditori sopra boschi*.[1] The title immediately suggests sustainability. *Provvedere* – as we already know from the Latin word *providentia* – means to care for, to look after, to provide for, to procure. It was no coincidence that the appointment of 'Providers for the Forests' happened just at this time. La Serenissima was at the height of her territorial powers and extension. The *stato da mar*, the 'Domains of the Sea', stretched to the eastern edges of the Aegean. The *terra ferma*, the mainland territory, covered the area from Ravenna to the Dolomites, from the Istrian peninsula to the north Italian lakes. But five years earlier, the Turks had conquered Constantinople and continued their expansion towards Greece and the Adriatic. Venice, caught up in costly wars with Genoa and the Habsburg Empire, was at the same time arming itself for an extended naval war in the eastern Mediterranean. In 1476 the Venetian Senate passed new laws regulating the use of mainland forests. They turned out to be "the most significant piece of forestry legislation ever passed by the Senate" and remained a model "until the fall of the Republic" (Karl Appuhn, in his brilliant study *A Forest on the Sea*). Their fame spread throughout Renaissance Europe.

The *provveditori sopra boschi* took charge of forestry and of the market in wood. In the following decades they purchased or requisitioned a series of hitherto locally-used wooded areas and turned them into state forests: Il Montello, oak woods north of Treviso, the Bosco del Cansiglio, an extensive stock of beech trees near Belluno, and the forest of St Marco, pine woods on the slopes of the Dolomites. All of these woods were not far from the banks of the Piave. Their timber could therefore be transported as far as the lagoon by timber rafting.

The influence of the *proti*, the governors of the Arsenal, is conspicuous in all of this. The demand for wood for the state shipyards, or more precisely their hierarchy of types of wood, inscribed itself into the forestry policy of la Serenissima and thence into nature. A plethora of prohibitions and regulations now limited access for those who had hitherto used the woods, the local cattle breeders, charcoal burners and craftsmen. The state power took control of nature and remodelled it. A natural world was created which corresponded to human priorities. Geometry aided this process. The forest inspectors brought with them surveyors and map-makers, who marked off the borders and drew up cadastres, or land registers. Tree species, and the age and quality of stocks, were recorded as accurately as possible. By around 1569, more or less every oak in the Venetian forests had been registered. Many tree trunks were stamped to indicate their future usage. On the basis of the surveys, procedures were developed for rotational felling, that is, for an ordered chronological sequence for the harvesting of the wood in a particular area. In the first inventory of the beech woods at Cansiglio in 1548, the three *proti* asserted that in their estimation the forest contained "beech timber of great size and excellent for making oars", and claimed the forest "is of such size. that it can be preserved within its present limits, unless it is wasted and ruined, as it is happening every day now by making charcoal and cutting firewood. So we believe we can promise your Excellencies that it will provide oars for your needs for many tens, no hundreds of years at a rate we would conservatively put at between six and ten thousand oars a year." Such is the sustainable time horizon of la Serenissima – planning for centuries. Ironically, the galleys for which the *provveditori sopra boschi* were concerned here to ensure a supply of oars had all gone to the scrapyards less than 200 years later.

Over the next few decades, the forestry authorities' investigations showed a steady increase in the reserves represented by the woods. Then came a turning point. From that point on, the statistics showed a continuous decline in the stock of exploitable wood. The fear of a coming shortage became a constant feature of the political discourse of Venice. It brought about a dramatic decision in 1630. When the Senate took a vow, in the middle of the great plague which claimed 50,000 lives in 1629-30, to build a huge new church as an act of penance, the decision was vetoed by the *provveditori sopra boschi*. At issue were the 12,000 oak trees required for the foundations. To fell this many trees at once would eat into the oak forest reserves, and would jeopardise shipbuilding and the reinforcement of the canal beds and breakwaters. What to do? To break a solemn vow was unthinkable. But to ruin the forests was likewise out of the question. The solution: Santa Maria della Salute, the baroque jewel at the mouth of the Grand Canal, was built on oak timbers which Venice bought in Hungary, at considerable additional cost. This is a rare and fascinating opportunity to see a mature society working out its priorities. Spiritual needs come in first place, and second comes the ethical principle of providing for the future (providence). All other considerations are subordinate. These priorities will be implemented – whatever it costs!

The design put forward for the basilica by the Master Builder Baldassare Longhena won the support of the Senate because it "achieved an extraordinary decorative impact without costing very much". The elegance of simplicity – this too is a lesson in sustainability. But how far removed from the spirituality of the Canticle of the Sun! Although the Franciscan order was well represented here, in this city on the lagoon where its founder had preached to the birds, next to nothing now remained of its original ethos.

Importing wood was not a lasting solution. Throughout the period of its decline, Venice never fully resolved the resource crisis in its wood supplies. What were the causes of this failure? The American historian Karl Appuhn sees the fateful mistake in too much centralisation and bureaucratisation of resource management. It is probabaly true that the Venetian authorities had arrogantly ignored the traditional wisdom and knowledge of local forestry, and the ecological intuition, of the people living in and around the woods, the woodcutters and suchlike. Instead they had imposed on the forests, in the interests of the Arsenal, an artificial, an unnatural system of rational resource management which could not be maintained over the longer term. But I have a suspicion that the key to understanding the Venetian resource crisis lies elsewhere.

Between 1576 and 1578, while the *provveditori sopra boschi* with their surveyors were mapping the woodlands bordering the Piave in order to manage their future exploitation, the famous Venetian painter Jacopo Tintoretto completed four works now hanging in the Anticollegio of the Doge's Palace. One of them shows the goddess Minerva protecting Pax, the goddess of peace, and Abundantia, the personification of abundance, against Mars, the god of war. The scene of the conflict is a hilltop overlooking a tranquil valley. Mars, in full armour and brandishing a lance, steps out of a dark background to threaten the three women. His purple commander's cloak has been laid aside on a rock. In the middle, pushing him back with an energetic movement of her arm, is Minerva. She too is martial. Her hips and upper arms are still covered by armour, her spear is to hand. But otherwise she is dressed only in a thin blue shift. With her right hand she shields the defenceless, almost naked Pax, who, wearing a laurel wreath, seeks refuge on the brighter side of the picture, which is lit by the dawn. Next to her, entering the picture from the left, is Abundantia. She bears the emblem of her power, the horn of plenty.

Here it is, this symbol which is so central to our civilisation. The cornucopia – in Greek mythology, the horn of the she-goat Amalthea. Zeus had accidentally broken it off while playing with the beautiful animal, and out of remorse he endowed it with magical powers, namely the ability to supply nectar and ambrosia in an unending stream and to fulfil all wishes.

A contemporary interpretation of Tintoretto's painting was that "it portrays the wisdom of the Republic, which understands how to keep the horrors of war at bay, so that the citizens may lead happy lives, and in enjoyment of their peaceful existence may learn to love all the more passionately the state which affords them it". With the perspective of history, the picture allows a different interpretation. The defence of *abundantia* – affluence – was the hidden core of Venetian state policy – and the cause of its failure. The demands for ever more luxury goods from a society built on the idea of the cornucopia, and addicted to affluence, will sooner or later come up against the limits to growth. Even under a thoroughly rational system of resource management, those demands will inevitably exceed the carrying capacity of the ecological systems. They are quite simply not sustainable.

'Il Montello', a 'site of memory', spring 2009. What has become of the famous oak forest of the Arsenal? North-west of Treviso, the Piave flows in a broad green sweep towards the Adriatic. The subsoil here is pale chalk and dolomite, the vegetation sparse, predominantly coppice. The hilly landscape is radiant. Solitary oaks and chestnuts rise up out of the undergrowth of whitethorn, acacia and hazelnut. There is much light green – the light trembles on the footpaths. Here and there a private villa, a vineyard, an orchard. The old system of the 'Prese', the twelve forest tracts into which the woods were divided, can still be discerned. A path of soft gravel cuts across the 'Presa' of the old oaks. From one end to the other it is only an hour's walk. Small groups of ancient trees still stand here – the last relics of a great history. Information boards have been put up along the path. One of them reveals that in 1668 the Venetian Senate prohibited anybody who lived in the area from crossing the Piave to enter the forest, on pain of death – a deterrence against the theft of wood. Was this the final legacy of the Venetian experiment in managed forestry? In that same year, the construction of Santa Maria della Salute was nearing completion. In fact, the forest was destroyed neither by the poor wretches who lived nearby, nor by the Arsenal. In the First World War, the battle front ran right through this area. 'Il Montello' burned and perished in the hellish crossfire of those battles.

" . . . to manage woods discreetly!" [2]

"I this day delivered my Discourse concerning Forest-trees to our Society upon occasion of certaine Queries sent us by the Commissioners of his Majesty's Navy." This is the entry under the date of 15 October 1662 in the diary of John Evelyn. This day marked the beginning of an ambitious project to establish a sustainability policy for the forests of England. The diarist was one of the most glamorous figures of his age: owner of great estates, a passionate landscape gardener, entrepreneur, beekeeper, collector of coins and engravings, author of bestsellers and adviser to the King, an elegant courtier and a pious man. The full title of his paper was *Sylva, or A Discourse of Forest-Trees and the Propagation of Timber in His Majesty's Dominions.* Only a few weeks earlier, the commanders of the

Royal Navy had addressed an urgent enquiry to the Royal Society. They wished to know how a looming shortage of timber could be prevented. For the armed forces, this was a life or death issue. Over the course of the 17th century, the English fleet had been continually expanded. At this time, it found itself in a drawn-out naval war with Holland which was bringing heavy losses.

One of the navy's highest commanders was Prince Rupert, a figure of truly European stature. The legendary 'mad cavalier' was the son of the Count Palatine of the Rhine Frederick V, the 'winter king', and his wife Elizabeth, the daughter of James, King of England, Scotland and Ireland. His father's ascension to the throne of Bohemia sparked off the Thirty Years' War. His sister Elisabeth was a friend and interlocutor of Descartes, and corresponded with Leibniz. His brother, the Elector Palatine, tried unsuccessfully to bring Spinoza to the newly-founded university at Heidelberg. Rupert himself began his military career at the battle of Lemgo in the 'German War', the Thirty Years' War, was captured by the imperial forces, fled to England (where he led the troops of his grandfather James I against Cromwell's rebels), commanded a pirate fleet in the Caribbean after the defeat of the Royalists, and became the first Governor of the Hudson's Bay Company in Canada. After the Restoration in London, he rose to Admiral of the Fleet and led the naval war against the Netherlands. He later became a Fellow of the Royal Society.

In this era, one warship after another was being launched from the shipyards. The construction of a medium-sized frigate alone consumed almost 3,000 mature oaks. The forests of the island kingdom were shrinking rapidly. The predicted crisis of resources and energy threatened not only the wooden 'bulwarks of this nation', the navy, but the whole society. The causes, in addition to shipbuilding, were principally the conversion of wooded land into farmland in order to feed a growing population, and the beginnings of industrialisation in glass manufacturing, iron forges and gunpowder production. Another problem was the abolition and abandonment during the republican regime under Cromwell of ancient laws and customs regulating the use of the forests. Experts predicted that 'in an age or two hence' the kingdom would experience an acute economic and social crisis caused by a shortage of timber and firewood.

The newly-established Royal Society had developed out of an interdisciplinary debating and experimenting circle centred on the chair of astronomy at Gresham College in London then held by Sir Christopher Wren. It had its rooms in Bishopsgate – not far from the Stock Exchange, where today the monumental office block that is Tower 42 looms over the City. Its Secretary was Henry, or Heinrich, Oldenburg, a German with a very wide network of contacts. He maintained a particularly long and intensive correspondence with Spinoza. Oldenburg asked four Members of the Royal Society to work on the enquiry from the navy concerning the shortage of wood. All four were 'virtuosi'. This term was applied in Europe at that time not only to gifted musicians, but to outstanding and innovative individuals from many different fields of knowledge and practice, universal minds with both a broad purview and highly developed specialised knowledge – liberated thinkers. They were frequently advisers at court, but not its creatures. They formed not closed circles but open networks of communication which spanned the whole of Europe and beyond. *"Omnia explorate, meliora retinete."*[3] Explore everything, keep the best. Evelyn's motto expresses the mentality of a virtuoso.

The four experts from the Royal Society were all born between 1606 and 1620. What qualified them for their task?

Dr Jonathan Goddard, doctor and physicist, founding member of the Royal Society, had applied himself to the anatomy of woody plants and to vegetative motion, that is, to plant growth. By this means he had gained insights into the sowing and planting of trees and the optimal times for harvesting wood, all of which went into Evelyn's *Discourse*.

Like Goddard, Dr Christopher Merret had a successful practice as a doctor to London's upper classes. His interests ranged widely. Because of his experiments with secondary fermentation, he is today considered the inventor of the *méthode champenoise*. Merret drew up the first list of native English birds, made contributions to the improvement of glass manufacture, and published a paper on tin mining in Cornwall. The suggestion that fossils were organic in origin was first made in one of his papers. Merret had gathered information on the practice of forestry in Germany and France which he now volunteered.

John Winthrop came from one of the famous English families of puritan emigrants to America. His father is considered a 'lost Founding Father' who in 1630, shortly before the emigrants' flagship *Arabella* first landed in the New World, gave the speech about the 'City on the Hill' which has shaped the American Dream ever since, with its assertion that the Puritan colonists were God's chosen people on whom the eyes of the world were now fixed. John Winthrop the Younger followed him to America, and served already as Governor of the colony of Connecticut in 1662. Two years later he took part in the expropriation by the English of the colony of 'Nieuw Nederlands' on the island of Manhattan from the Dutch. He had already introduced to the Royal Society a plant cultivated by the native Americans, maize, which was unknown at that time in Europe. Winthrop also established the first ironworks in New England and experimented with the production of tar and pitch distilled from the resinous wood of the American pine. These materials were at that time indispensable for the waterproofing of ships' timbers. Winthrop contributed to the discussions at the Royal Society his knowledge of the virgin forests of America.

But the head of the small commission was John Evelyn. On that day, the 15th October 1662, he gave the main presentation and was given the task of working up all of the contributions into a book. He was then 42 years old. He set to work, and a little over a year later, in February 1664, the book appeared, in a run of 1,000 copies. It was the first publication by the Royal Society. Its title was *Sylva* – the Latin word for woods.

John Evelyn proudly described himself as being "wood-born". His parents' estate, Wotton Place, a Tudor manor house surrounded by an extensive park, lay in the wooded valley of a stream in the county of Surrey south of London, on the lower slopes of a range of low hills which the Anglo-Saxon invaders 1,000 years earlier had named simply 'The Weald', or in modern German 'Wald' – the forest. This area is regarded as the cradle of the British iron industry. Evelyn's ancestors had made their fortune from gunpowder. In addition to brimstone (sulphur) and saltpetre, the other raw material for the powder mills where the explosive was produced was charcoal, which was burnt in the kilns in the local woods. Evelyn's father, "a thriving, Neate, silent and methodical genius", maintained several hammer mills and wire-drawing mills for working iron and copper.

However, the young Evelyn's great passion was garden design. Most of his twenties he spent on the Continent, partly in order to acquire a rounded humanistic education on the 'Grand Tour' (obligatory for young men of the upper classes), partly in order to escape the bloody chaos of the civil war between the English monarchy and Parliament. He spent some time in Holland, and then lived, travelled and studied in France and Italy. His diary of these footloose years is full of enthusiastic descriptions of the gardens and parks which he visited assiduously. Among them was of course the famous Hortus, the botanical garden of the Dutch university of Leiden. In Paris he was captivated by Cardinal Richelieu's garden. And the Jardin du Luxembourg, already at that time a public park, seemed to him a "paradise". He admired the geometric arrangement of the roundels and walkways, the strict symmetry of the flowerbeds, the ingenious sightlines and undeviating avenues of trees, the carefully placed water fountains. But he also appreciated the plots of "natural wildness" embedded within even these highly artificial French gardens.

The years 1644 and 1645, when he was 24 years old, he spent in Italy. Like most educational tourists of his day he stood in awe on the crater of Vesuvius and meditated on the raw power of Nature before making a pilgrimage to the grotto containing the grave of Virgil on the other side of Naples. In his bucolic verses the Roman poet celebrated the rural idyll, and in his *Georgics* he produced what remained the standard textbook on farming and forestry up to the Renaissance. *Sylva* opens with a quotation from Virgil, and he is the most frequently quoted author in a book which is peppered with classical citations. Evelyn is fascinated by the Renaissance gardens of Rome, the evergreen Mediterranean tree groups of the Villa Borghese, the cypress grove at the Villa Medici, the terraces and sightlines of Tivoli, the "romantic" waterfalls and grottos in the parks of Frascati. For the final part of his sojourn in Italy, he registered at the university of Padua to study physics and anatomy for a couple of months. It was probably at this time that he went to see the woods on the banks of the river Piave, which he described in *Sylva* as "a goodly forest of Oaks, preserved as a Jewel, for the only use of the arsenal, called the Montello".

In 1652 John Evelyn returned to England and settled in Sayes Court in Deptford. The estate, which had been left to him by his wife's parents, lay six miles downriver from London Bridge along the Thames, on the road to Greenwich. It bordered the huge naval shipyards. At this spot, close to the river and open to wind, weather and – even in those days already – the London smog, he began to lay out a park which would reflect his ideas on garden design.

An avenue of lime trees led from the road to the house. Immaculately straight hedges of holly or elder divided the space into geometrical forms. Elms planted in rows of military regularity marked the external boundaries. The native tree species were gathered in a grove which offered a spot for contemplation and communication. An orchard of mixed fruit trees and a herb garden served to supply the kitchen. A "perfect oval, planted with cypress trees", at the centre of which two promenades crossed at right angles, formed the heart of the garden. It was both "ornament" and "magazine" – the combination of the beautiful with the useful. His vision was of an "Elysium Britannicum", where the ancient dream of the "isle of the blest" would be realised in the English countryside, a reconstruction of the garden of Eden. "Paradise" – or in Hebrew, "pardes" – originally meant "garden of trees".

The exact observation of nature, endless pleasure and spiritual experience were by no means mutually exclusive during that period of the early Enlightenment. Evelyn "never tired of celebrating the mystery by which the tiny grain, 'which lately a single Ant would easily have born into his little Cavern', with its 'insensible rudiment, or rather habituous spirit', would ascend 'by little and little . . . into an hard erect stem of comely dimensions, into a solid Tower as it were' ".[4] For the enlightened minds of his time, wonder – *thaumazein* in ancient Greek philosophy – was a component of a proper and fulfilled life. It was the readiness and the capacity to perceive and receive the world with open eyes, with trust and with a constantly recurring sense of wonder.

Evelyn's private gardens developed within a few years into a very popular London attraction, visited with pleasure even by the royal family. *Sylva* was written here, in this beautiful setting near the naval dockyards. The book's response to the resource crisis certainly went well beyond the Royal Navy's horizons.

Much more than a memorandum on resource management for the naval dock-yards, *Sylva* is a handbook on forestry for the English landed gentry. And beyond that it contains the germ of an aesthetics of the natural world and of a guide for ecological living. Apt quotations are scattered throughout – from the Bible, from Greek and Roman authors such as Aristotle, Virgil, Pliny and Ovid, from later authors such as Nicholas of Cusa and Shakespeare. The book describes around 80 native and foreign tree species, their botanical and aesthetic features, their respective specific uses. *Sylva* gives the reader precise instructions on how and when trees should be planted and transplanted, pruned and felled. Throughout, its aim is "to increase the beauty of Forests and value of Timber".

Where Evelyn describes the economic and social drivers of "the furious dev-astation of so many excellent Woods and Forests", his book reads like a modern UN Report on the destruction of the tropical rainforests. "The late increase of shipping, the multiplication of Glass-Works, Iron-Furnaces and the like," above all though "the disproportionate spreading of Tillage" and "the destructive razing and converting of Woods to Pasture" have brought about the annihilation of "the greatest magazines of the wealth and glory of this nation". And "this devastation is now become . . . epidemical." The root causes? Evelyn attacks head-on those of his contemporaries "so miserably lost in their speculations" that they seek only "to satisfie an impious and unworthy Avarice".

"May such woods as do yet remain intire be carefully Preserv'd and such as are destroy'd sedulously Repair'd". The key term of the book is *preserve*, or alter-natively *conserve*. The *conservare* of the old doctrine of Providentia appears here in a new context, and now as a duty of the whole society.

Can a natural resource be intensively exploited over a long time and at the same time substantially preserved? This question, still central to every sustain-ability strategy today, is Evelyn's topic. He gives an indication of a possible answer when writing about his father's iron foundry in the woods at Wotton Place in Surrey. "I have heard my own Father . . . affirm, that a Forge, and some other Mills, to which he furnish'd much Fuel, were a means of maintaining, and increasing his Woods; I suppose by increasing the Industry of planting, and care . . . This may appear a Paradox," Evelyn writes.[5] But his father achieved precisely this aim. The woods in their current condition were "a most laudable Monument of his Industry, and rare Example." And he continues: "Without such an exam-ple," that is, where natural resources are not handled in a sustainable way, "I am no advocate for ironworks, but a declared Denouncer." Presumably encouraged by John Winthrop, he recommends the transplantation of the iron industry from Old England into the densely-wooded territories of New England. "'Twere better to purchase all our Iron out of America, than thus to exhaust our Woods at home." For these are only "inexhaustible magazines" if they are handled "with care". Evelyn's formulation for this is "to manage Woods discreetly". This means to differentiate between individual forests and to manage them carefully in accordance with their respective ecological characteristics.

To be sure, Evelyn saw plantation as the best way to increase the stock of wood. Sowing and planting seemed to him the most reliable means of solving the resource crisis. The trees should grow in plantations, in nurseries, protected by fences in the first few years from the browsing of livestock and wild animals. "They may be sown promiscuously, which is the most natural and Rural, or in straight and even lines" – in geometrical order. Certainly, tree-planting has to take into account the characteristics of the habitat. "But first it will be requisite to agree upon the Species; as what Trees are likely to be of greatest Use, and the fittest to be cultivated." Like Charles Darwin 200 years later, Evelyn uses here the word "fittest" to denote those which are best adapted. He advocates "speedy growing" species, specifically spruce, which were comparatively rare in England, and the trunks of which would make excellent ships' masts. But "it is not till a Tree is arrive'd to his perfect Age, and full vigour, that the Lord of the Forest should consult, or determine concerning a Felling. For there is certainly in Trees . . . a time of Increment, or growth; a Status or season when they are at best (which is also that of a Felling) and a decrement or period when they decay."[6] Evelyn is convinced that the human mind can impose a new order on wild nature. And that – not least in the interests of the coming generations – it must do so.

"Let us arise then and plant, and not give it over until we have repaired the havoc . . ." He buttresses this passionate plea with numerous examples of good practice drawn from all over Europe. He points not only to the woods at Montello but also to "the Noble Forest of Nuremberg" as an example of "an almost continual forest". He reports with approval on the legislation in Luxembourg which bans "any farmer from felling a tree unless he can prove that he has planted another".

When he describes the practices of the owners in France and Germany who "divide the Woods, and Forests, into eighty partitions; every year felling one of the divisions: so as no Wood is fell'd in less than fourscore years", Evelyn touches upon the simple organisational rules from which sustainable forestry has evolved.

His most urgent plea – and the leitmotif of his book – concerns provision for the future, for posterity. He quotes a Latin saying: each generation is *non sibi solus natus* – not born for itself alone, but rather born for posterity.[7] But, he adds accusingly, his own contemporaries were apparently *fruges consumeri nati* – born to consume the fruits of the Earth.

At this point, Evelyn develops an ethics for a provident and responsible society. "I do not pretend that a man (who has occasion for Timber) is obliged to attend so many ages ere he fell his Trees; but I do by this infer, how highly necessary it were, that men should perpetually be planting; that so posterity might have Trees for their service of competent, that is of middle growth and age, which it is impossible they should have, if we thus continue to destroy our Woods, without this providential planting in their stead, and felling what we do cut down with great discretion, and regard of the future."[8]

"Providential planting" is a highly interesting 'recycling' of the ancient discourse on *providentia*. Good husbandry is the central concept of Evelyn's underlying economic theory. This traditional term describes the careful, efficient, truly "economical" landowner's use of the available resources. It transcends a purely managerial approach to nature. For what is the secret of good husbandry? "To obey nature."[9] And what is its goal? "To render the countries habitable." We are very close here to the vocabulary of our modern sustainability discourse.

Sylva was a huge success. The planting of trees became a national sport for the landed gentry throughout Britain. Nevertheless, the book had no lasting impact on the practice of forestry, nor on the economy as a whole. A combination of two strategies seemed to Evelyn's contemporaries a more secure route to never-ending prosperity: 1) to import natural resources that were becoming scarce at home from around the entire globe; and 2) to substitute other raw materials for any threatened by shortages. Free trade and colonialism, plus technical innovation, became the key to a future at whose end we have arrived today.

This dual strategy is exemplified in the reconstruction of the City of London after the Great Fire of 1666. It was relatively easily accomplished using imported timber from the forests of Norway and from the American colonies, and by substituting brick and stone for much of the wood which had been the main construction material before the fire. (Incidentally, it was Evelyn who drew up one of the best-known plans for wholescale state-led redesign of the city – ignored, like all the others, in favour of piecemeal rebuilding on the existing medieval street map.) The replacement of wood as an energy source by fossil fuels was already well under way. Coal, at first the 'sea coal' transported by ship from the area around Newcastle, appeared to be able to meet comfortably the demand for energy from rapidly growing industries and a growing population. We shall return to this.

Evelyn's plea for responsible management of renewable resources fell on deaf ears. After his death in 1707, the book continued to be read as an elegant piece of literature for lovers of the art of gardening. Only the admirals of the navy continued to make occasional warnings about wood shortages. Their calls finally fell silent in 1862, when a sea-battle between two ironclad warships in the American Civil War, the battle of Hampton Roads, brought an end to the age of wooden ships.

The forestry reforms of the Sun King

In the year 1664, when *Sylva* was the talk of the town in London, a book of fables was published in Paris. One of them tells the story of an 80-year-old man who plants a tree in front of his house. The watching neighbours' children mock him. To build a house would be understandable, but to plant a tree? At your age? What fruit could your labours possibly still bring you? Why burden your life with caring about a future in which you will not share? Think rather on the mistakes of your past and take leave of long-term hopes and plans. The old man replies: "Every new enterprise comes too late. None endures. In any given instant, we cannot be sure even of the next. But my descendants will thank me for the shade given by this tree. So allow this old man the pleasure of caring for the wellbeing of others. This is a fruit which I can already enjoy today."

'The old man and the three youths' is a light-hearted fable, and a parable about a key question of sustainability. What do I gain from doing something for coming generations? The answer: a deep satisfaction!

The fable is from the pen of the French writer Jean de La Fontaine. At that time, when he was not staying in Paris, the author lived in his home town of Château-Thierry, a good 50 kilometres west of the capital, in Picardy. His main occupation was as a lawyer. Like his father before him, he held the post of 'Maître des Eaux et Forêts'. So the story-teller was responsible for the forests and the fishing waters of his native region. In France, too, planting trees was now more than a hobby for old men. It had become a matter of supreme national interest.

In the year 1669, after years of preparatory work on the part of his civil servants, Louis XIV issued the 'Ordonnances sur le fait des Eaux et Forêts', the Ordinances of the Waters and Forests.[10] He signed the 200-page document on the 13th August of that year at Saint-Germain-en-Laye. The royal castle lay at the edge of an extensive wooded area in a loop of the Seine a couple of kilometres west of Paris. It was Louis' birthplace, and at that time still his main residence, though the conversion of the former hunting lodge at nearby Versailles into a palace of unprecedented dimensions had already begun. Tens of thousands of workers laboured at the site. Hundreds died a wretched death. The park was laid out on sand dunes – a triumph of geometry over nature. The king's greatest pleasure, a critic remarked, is to tyrannise nature.

Louis XIV was 30 years old when he signed the new laws governing forestry. The preamble sets out the aims in flowery language: to repair the apparently "almost irremediable, universal and inveterate disorder" into which the woods and waters have fallen; to "bring into bloom once more this noble and precious part of our domains"; to enable it to produce once more "in abundance all the various public benefits which may be anticipated". "Good and wise regulations" would now establish and ensure once again that "the fruits will be passed on to posterity." *Faire passer le fruit à la posterité* – another blueprint for the Brundtland formulation. From the court of the Sun King![11]

The work on the forestry reforms had begun eight years earlier, in the summer of 1661. Louis had just taken over absolute control of state policy and of the life of the country. For his emblem he chose the Sun. The first two decades are regarded as the most innovative and productive phase of his long rule, and the forestry reform as the centrepiece of a wide-ranging modernisation drive. It bore the signature of the recently appointed Controller-General of Finances and councillor of state, Jean-Baptiste Colbert. The most powerful of Louis XIV's 'doméstiques', Colbert was a pedantic bureaucrat, a servile and greedy creature of the court, and – at the same time – a far-sighted moderniser. He followed a sweeping plan for the enhancement of the *gloire*, the fame of the king and thus of France. Colbert's vision was not that of the rural *Elysium* which John Evelyn imagined for England; nor the humane, provident thinking of La Fontaine's old man. Rather, he planned to transform France into a huge factory, and the first industrialised nation in Europe.

For that he needed money – and timber. The new factories were dependent for their energy supply on cheap charcoal. Similarly a continuous growth in external trade, a key factor in Colbert's mercantilist economic plan, was dependent on wood. Trade was regarded as war by other means. It was not possible without a merchant navy. The protection of one's own ships and the elimination of competition from other trading nations both necessitated an armada of warships. In 1661, the French navy was as good as non-existent, whereas the great rivals – England and, even more successfully, Holland – had built up powerful naval forces. Shipbuilding became an obsession for Louis and his Minister. However, a secure, reliable, long-term and plentiful supply of timber and tar for the new shipyards, and the energy supplies for the factories, required – as in England – the solution of a strategic problem: the depredation of the nation's forests had to be reversed, by all and any means.

"La France perira faute de bois" – France will perish for lack of wood. With this shrill alarm, Colbert introduced the forestry reforms. He identified their three purposes, in descending order, as: to restore the income to the treasury from the royal forests; to dispel the fear of a coming timber shortage; to ensure enough wood for the shipbuilding industry. The overarching idea had been formulated by the Sun King himself in a handwritten note: ". . . il était nécessaire de faire un bon ménage des bois" – the need for good management of the woods.

The forestry reforms began like an ambush. In November 1661, the authorities cut off access to the royal forests, suspended the traditional rights of those living within or nearby, and stopped the sale of wood. A thorough inventory was carried out on all parts of the vast area of wooded land belonging to the crown, which stretched like a patchwork across the country from the deciduous forests of the Ardennes to the pine groves of Provence and from the Normandy forests to the mountain groves of the Pyrenees. Specially-assembled squads of loyal civil servants swarmed out into the provinces and, bypassing the regional Grands Maîtres des Eaux et Forêts, began investigating the condition of the forests and the uses to which their wood was put.

Their reports define the problem in political and legal terms. The subjects' traditional claims to the enjoyment of the woods conflict with the long-term interests of the state. However, their claims have no legal basis. What is currently happening in the woods is quite simply criminal. The woods belonging to the crown are being unscrupulously plundered and over-exploited by speculators, wood merchants, and by the lords of the landed aristocracy just as much as by the farmers, cattle herders, vagabonds and landless paupers. Theft, misuse, arrogation of rights – the ruination of the woods is presented as the consequence of everyday criminality on the part of subjects from all classes of the population.

This interpretation moulded the ordinances of 1669. They are in essence a bundle of measures to restore the power of the government over the royal woods and to secure control. They even regulate the breadth and depth of the ditches separating the state forests from private wooded land. People living in the woods are forbidden to carry out wood-based trades. Anybody found carrying an axe or a saw off-track in the woods at night is threatened with imprisonment. Vagabonds and 'idlers' are forbidden to stop in the vicinity of the state forests. Lighting a fire in the woods is strictly punished. The forest wardens are to be armed with pistols. Other paragraphs prescribe a minimum age for forestry officials and fix drastic penalties for arson. The area given over to woodland grazing is reduced and the selling market for wood is reorganised.

In 1666, Colbert sends one of his best foresters to the Midi, the south. Louis de Froidour begins his investigations from Toulouse. It is thanks to him that the ordinances are in the end more than just a catalogue of bureaucratic and legalistic regulations. For now the forestry aspects move to the fore. The southern forests are in dire condition, Froidour reports as well. But he differentiates between *taillis* (coppice or low forest) and *futai* (timber or old-growth forest). Systematic harvesting is unknown. Young trees are not given time to mature. The crisis of the forests is primarily a crisis of old-growth forests. The coming shortage will be a shortage of mature timber.

These insights feed into the ordinances. When an area is harvested, a certain number of mature trees must be left to provide for seedlings. Bare patches, that is, clearcuts or clearings, must be reforested by sowing and planting. One quarter of every area of coppice forest must be set aside and allowed to grow into timber forest. The French verb *retenir*, meaning 'retain' or 'set aside', i.e. reserves, is a key word in these documents. *Conservation* is another one.

La conservation des bois does not mean conservation or protection of the woods in the sense of not utilising them. The emphasis is rather on maintaining – sustaining – the productivity of the woods, protecting their ability to regenerate, to replenish themselves, and thus their capacity to produce wood forever – *à perpetuité*. *Conservation* signifies *sustaining use*, which requires making the replenishment of natural resources a condition of their use.

The text of the *Ordonnances* speaks of the *aménagement de nos forêts*, the management of our woods. Another key term, one we have already come across in Evelyn's *Sylva. Aménagement* has a very interesting etymology. Its root is the Latin *mansio* – a housing, a place to stay, an accommodation, or in medieval Latin also an estate, a farmstead, a settlement. The French *maison* and the English *mansion* developed from this root.

So, what exactly is *l'aménagement*, or management? In modern French, *aménager* is to furnish a home. Originally it meant: to bring something *ad mansionem*. That is, to prepare natural resources and transport them to one's home, the place where they are used. The idea of making nature fit human needs is inherent in the old usage. The concept of *gubernatio* from the doctrine of Providentia is also resonating within this usage, the idea of directing or steering nature; as is the idea of good husbandry, of the economical use of resources. The king himself had spoken of the *bon ménage des bois*. The *Ordonnances* distinguished between good and bad *exploitation*. The latter means squandering, depletion; that is, using resources at a level or rate which exceeds their capacity to regenerate.

Management consists essentially of regulations. *Régler* and *règlement* are other key terms in the text of the *Ordonnances*. Their underlying message is that the wood harvest should be regulated in a way which ensures that the use of the woods is brought into line with their capacity. To this end, they prescribe a strict *règlement* of all activities which impact on the woods. The analysis which lies behind this is that if their use continues to be determined by the needs, desires and free will of the local residents or the society at large, then the wood shortage will inevitably intensify. *Laisser faire* will lead in the long term to collapse.

The Ordinance, which came into effect in 1669, had some initial success. Within ten years the royal income from the sale of wood had risen substantially. So this aim at least was met. However, the triumph was short-lived. The application of the new regulations proved to be a truly Sisyphean task. Parochial local interests proved tenacious. Over the longer term, the 'Grande Réformation' of the forests achieved very little. On the eve of the Revolution in 1789, there was less woodland in France than in 1669.

But under the pressure exerted by the timber shortage, and under the umbrella of what might be termed an early attempt at a national sustainability strategy, natural science blossomed in 18th-century France. People wanted to uncover the secrets of nature, of its productivity and fertility, in order to secure their own wellbeing and that of future generations. An example is the Comte de Buffon, the great French naturalist. John Evelyn's *Sylva* was a formative reading experience of his younger days. He became the curator of the royal botanical gardens in Paris, author of the monumental *Histoire naturelle* and a great literary stylist. During most of those years he managed a forest and an ironworks in Burgundy. It can still be visited today, having been painstakingly restored, a couple of kilometres outside the small town of Montbard.

From Colbert's *Ordonnances* it was concepts such as *coupe reglé, conservation, aménagement* and *bon ménage* which had the longest afterlife. They paved the way for the 20th-century concepts related to sustainability. However, they are marked by a Cartesian view of nature as an object to be subjugated – the managerial perspective.

The emerging sustainability theory still lacked a central concept. There was no single word which could concisely express the basic idea. The word which would do this was coined in Germany. When their little worlds collapsed during the Thirty Years' War, the economists and foresters of the German mini-states developed a vision: the possibility of providing a long-term, secure and steady supply of wood from *der ewige Wald* (the eternal forest). Out of this vision emerged the concept of *Nachhaltigkeit*, sustainability. A journey in search of its roots takes us back where 200 years earlier the silver rush had inspired the horror story of the 'Murder of Mother Earth'. The 'site of memory' this time is Freiberg, the 'silver town' in the foothills of the Ore Mountains.

CHAPTER SEVEN

The creation of the word

On one side of an alley leading in a long curve from the castle to the cathedral, a massive, late-Gothic building rises high above its neighbours in the unbroken row of housefronts. For over 350 years its thick walls have enclosed the offices of the Chief Mines Inspectorate for Saxony in Freiberg. The squat doorway is flanked by two stone benches and protected by a canopy. The visitor steps inside through a heavy wooden door, and once inside the entrance hall is transported to another era. The low, stellar-vaulted ceiling, supported by a single column and now plastered white, was once painted in blue and gold, the colours of the sky. Footsteps echo on the heavy flagstones. The window recesses reveal the thickness of the walls. A hint of Faust, of the Renaissance, lingers in the air, the time of the seekers after truth, the alchemists and diviners. At the time the Mines Inspectorate was founded, the search for the *lapis philosophorum*, the philosopher's stone, had by no means been abandoned.

A spiral staircase leads from the ground floor down to a vaulted cellar. There, one comes up against a bricked-up doorway. This was once an entrance into the underground labyrinth of tunnels and gangways, galleries and shafts of the Freiberg mines. Today these deep workings have been closed down and sealed off, left to decay in the silence. *Toter Mann* – 'dead man' – is what the local miners call a disused gallery.

An 18th-century Silicon Valley

In the late 17th century, the mining industry of the Ore Mountains was recovering from a protracted crisis. At first, it was the competition from the silver mines at Potosí which had put it under pressure. Then came the Thirty Years' War, which raged even more murderously here on the border to Bohemia than elsewhere. At the end of the period, countless mines lay in ruins and entire villages were extinct. The comparatively swift recovery which followed is considered one of the success stories of the history of Saxony. It was based on silver.

The mining district of the Ore Mountains was administered by the Mines Inspectorate like a state within a state. The ore is brought up into the daylight out of hundreds of mines: silver and copper, tin and cobalt. The metals are extracted at countless smelting plants. Nearly 10,000 miners toil day and night, bent double, squeezed into narrow tunnels, doing their bloody, bitter work with hammer and chisel, drill and gunpowder. In low separating chambers choked with dust and poisonous with arsenic, children as young as eight break apart the chunks of ore, driven on all the while by overseers with truncheons or straps. Muscle power is a principal source of energy. Every week, the 'silver coach', escorted by armed miners, transports the ingots to the mint at Dresden and into August the Strong's treasure chambers. The Ore Mountains silver finances the blossoming of the Residence into its full Baroque glory, the takeover of the Polish crown and the Great Northern War. In the years around 1700 the annual yield from the mines climbs slowly but surely to over five tons. Once again, the Ore Mountains form one of the most significant mining regions in Europe.

The office in Freiberg is the nerve centre of a scientific-technical revolution. In the era of the early Enlightenment, "knowledge, hard-won from experience" becomes "mistress of all government". But in this time of change, superstition – that mixture of mistaken belief and esoteric knowledge – is still alive. Hardly anyone dares to categorically exclude the possibility of the existence of mountain trolls. The mine authorities continue to employ sworn-in 'dowsers', who search for the subterranean veins of metals with their divining-rods. The prevailing belief is that metals are formed in the earth from a combination of brimstone (sulphur) and quicksilver (mercury). Even the most sceptical of natural scientists believes that the transmutation of metals may be possible. (Newton was still working on this around 1700.) At the same time, August the Strong is investing huge sums in experiments conducted by dubious gold-makers, who hope to create an artificial Eldorado in the secret laboratories of Dresden. But step by step the divining rod is giving way to efficient techniques of prospecting based on the science of geology. Alchemy is developing into chemistry. The Saxon Mines Inspectorate plays a substantial role in this process of modernisation.

They think globally here. The Russian Tsar Peter I pays a visit to the mines in 1698 on the way back from a study trip to Amsterdam and London on shipbuilding techniques. In 1711 he returns, takes a trip down a mineshaft, even takes hammer and chisel into his own hands, and is fêted that evening with music and a miners' parade in front of Freiberg Castle. The next day he meets the philosopher Leibniz in Torgau. Later, Peter I would bring in miners from the region into Russia to develop the mining industry there.

In the Mines Inspectorate the yields from the mines at Potosí are carefully noted down. The *Little Book of Mining*, written by the Spanish-Peruvian priest Alvaro Alonso Barba and published in German in Hamburg in 1676, is studied here. It describes the "famous mines of the Indian kings", situated 5,000 metres up on a bald and bare mountain-top, and tells of the "birth of metals" and of "Naphtha", also called "stone-oil" or "petroleum". For his part, Barba genuflects in his book before Jorge Agricola, the famous Saxon mining expert, and praises his "inextinguishable fame in the firmament of the sages of the Renaissance".

It is very probable that all available information on the *Arcanum*, the secret of Venetian glass-blowing, is gathered together here. It is even more likely that the sparse news emerging from Jingdezhen in southern China is followed very carefully. This is the global centre of porcelain manufacture, now also supplying the European market – at fabulous prices. The Mines Inspectorate is closely involved in the strenuous efforts being made in Saxony to discover the secret of making porcelain. The kaolin find in the St Andreas mine in Schneeberg in 1700, which provided a supply of the long sought-after raw material for porcelain, was directed from here. In 1705, the Inspectorate makes available five workers from the iron mills to support the research of the naturalist Ehrenfried Walther von Tschirnhaus and the alchemist Johannes Böttger in their secret laboratories in Dresden and Meissen.

" . . . a pure Spinozist"

The chambers of the Director of the Mines Inspectorate are still situated today on the second storey. From 1668 until his death in 1711, this was where Abraham von Schönberg carried out his duties. As the scion of an ancient aristocratic Saxon family which has held this post for generations, he learns both mining and the iron industry from the bottom up. He suffers from a "weakness of the thigh muscles", a partial lameness brought on by spending too much time as a young man wreathed in "mine-gases" and the smoke of the ironworks. But he rules his empire with an iron hand and an inventive spirit. He introduces the use of gunpowder for underground ore extraction. Mine surveying – the craft of measuring the physical extension of deposits and mine workings – develops notably, and the smelting plants are taken over by the state and centralised. Output rises slowly but steadily. Schönberg himself writes a handbook of mining, the first since Agricola's *De re metallica*.

His closest colleague is a metallurgist trained in France, Gottfried Pabst von Ohain. Ohain refines the technique for determining the metal content of ore samples. In 1701 the Saxon government persuades him to come to Dresden to provide support to the gold-makers in their experiments. Like his supervisor Schönberg, Ohain considers the transmutation of metals to be "against the laws of nature", and probably urged the abandonment of these senseless experiments.

But the idea of using the infrastructure set up for Böttger's secret investigations for the experiments in ceramics came from Ehrenfried Walther von Tschirnhaus.

Tschirnhaus is a mathematician, naturalist, inventor and philosopher of European rank. On his family estate he conducts experiments in optics and mechanics, and notably in solar energy. He constructs mirrors over a metre high out of concave copper plates which focus the Sun's rays onto a single point capable of burning diamonds and melting asbestos. He sets up a glassworks in Dresden whose furnaces require far less wood than conventional ones. He has for some time been following closely the Europe-wide efforts to produce porcelain, and has visited all the relevant workshops between Delft and Venice. His ultimate dream, however, is to track down the *ars inveniendi*, the art of invention itself.

This famous naturalist works in close cooperation with the Mines Inspectorate. Schönberg and Ohain are members of the small circle which Tschirnhaus gathers around him in the 1690s. What is planned is a society for Saxony along the lines of the English Royal Society and the Académie Française. As a young man already Tschirnhaus has played a role in the communications networks of the European Enlightenment. In the 1670s, his 'grand tour' takes him right through western Europe. In Leiden he studies medicine and natural science. In London he makes friends with Heinrich Oldenburg, Secretary of the Royal Society when Evelyn held his talk there on the timber shortage. In Paris, Colbert personally engages him as tutor to his son in Latin and mathematics. When Leibniz visits Paris in 1675, Tschirnhaus introduces him to Spinoza's *Ethics*, of which he has received a personal advance copy from the author. Tschirnhaus is already a member of the great Dutch thinker's innermost circle, his interlocutor and correspondent. When Spinoza dies in 1677, the Saxon aristocrat is one of the two executors of his estate.

However, until well into the 18th century the suspicion of Spinozism spelt the certain end of any scholarly career. When the young and ambitious philosopher Christian Wolff debates with Tschirnhaus in Leipzig in 1705 and subsequently writes of him that "he is a pure Spinozist", Tschirnhaus goes on the attack. It is not at all the case, he maintains, that Spinoza "confounded God and Nature with one another . . . rather, he defined God as *multo significantius*" – as the prime creative power.

Spinoza's ecological ethics are ever-present between the lines of Tschirnhaus's major philosophical work, *Medicina mentis*, published in 1695. "For Nature's treasures are inexhaustible," it tells us, and "God's majesty is visible in the small things as in the great." A founding principle of economics follows from this: that every benefit which man can obtain derives solely from the power of Nature. Tschirnhaus formulates a doctrine of happiness based on this insight. "The path to the attainment of the greatest happiness possible in this life" leads by way of "love of that which sustains my being or which serves my preservation". Is this new, Spinozist way of thinking inscribed into the new concept of sustainability?

A Saxon European

From this group of gifted administrators, scientists and engineers in the service of the Saxon mining authorities rose the man who wrote out the birth certificate for the term *Nachhaltigkeit* – and thus the blueprint for 'sustainability'. Hans Carl von Carlowitz formally became the number two in the hierarchy with his appointment as Deputy Director of the Inspectorate at the age of 32 in 1677. On Schönberg's death in 1711, August the Strong named Carlowitz as his successor. But his name appears only infrequently in the records. He seems to have played little part either in the practical administration of the mines or in the reinvention of porcelain. It seems that Carlowitz was appointed purely and simply to solve one key problem for the survival of the mining industry: wood supply.

Constructing the support frameworks for the shafts and galleries, and – even more – fuelling the charcoal-fired furnaces of the smelting plants and hammer mills, devour vast quantities of the precious resource. The environs of Freiberg and of the other mining towns are already deforested through continuous over-exploitation over the course of centuries. Logging ever more distant forests and transporting the timber by rafts can only provide temporary relief.

Many technical innovations of the time have their origin in this resource crisis: the replacement of traditional fire-setting by explosives as a means of loosening the ore, starting around 1643, serves to save wood. And the centralisation and complete nationalisation of the iron industry, through the establishment of a General Smelting Office in 1710, is explained in part by the hope of achieving greater efficiencies in the use of the dwindling wood supplies.

But despite all this, the price of wood rose unremittingly over the second half of the 17th century. A decree of the year 1677 reveals how ensuring the supply of charcoal was already perceived as the biggest problem for the Ore Mountains mining industry. Some hammer mills are already forced out of business by rising energy costs around 1700. When August the Strong undertakes a wide-ranging tour of inspection in 1708, the Mines Inspectorate declares "the unrestricted supply of the necessary wood" to be the central problem. It is at this point at the latest that Carlowitz is asked to work out proposals for a permanent solution.

A portrait from this period shows him as a baroque nobleman. It is in three-quarter profile and mounted in a medallion. His brow is marked with deep vertical lines. The thin-lipped mouth gives an impression of energy; his expression is serious and questioning. The dark ringlets of his long French wig fall onto the metal of his decorative suit of armour, over which he has thrown a velvet cloak. A light cloth is wrapped around his neck. The portrait is rounded off with the family coat of arms: an aristocrat and a 'virtuoso'.

Carlowitz benefited from an inheritance rich in practical knowledge. Hunting and forestry in the Ore Mountains had been the domain of his family over many generations. His father served as Overseer of the rafts and Master of the Hunt to the Elector. Carlowitz was born in 1645, in the last throes of the Thirty Years' War, in Schloss Rabenstein, a forbidding fortress which still stands today in the wooded hills west of Chemnitz. He grew up moving to and fro between the rough working life of the huntsmen, rafters and charcoal burners and the courtly life on the noble estates. He probably learned at an early age how to fire a charcoal kiln up to 1,000 degrees, how to steer a raft through rapids, or how to lay out a tree nursery. But he was just as assiduous in mastering the intellectual cosmos of his era. In 1664 he studied law for two semesters at the University of Jena (one year after Leibniz). Then he set off on his 'grand tour', a five-year journey of experience and study across the whole of Europe.

He is in Leiden while Spinoza is working on his 'Theologico-Political Treatise' in nearby Voorburg. In London in September 1666 he lives through the devastating fire which reduces the City of London to ash and rubble. This event was perceived as an apocalyptic catastrophe and compared with the destruction of Sodom and Gomorrah. Arson by French or Dutch spies was suspected, and Carlowitz became a victim of the ensuing hysteria. He is "attacked and beaten for a foreigner by the angry mob and thrown into a wretched gaol . . . ". He claimed later that it was only through the personal intervention of Prince Rupert that he was set free. He seems to have still been in the city, and may have witnessed the event himself, when in June 1667 the Dutch fleet sailed up the Thames and sank virtually the entire English navy at anchor at Chatham. Both the burning of the city and the sinking of the warships exacerbated the timber shortage and thus served to bring John Evelyn's book, which had appeared two years earlier, to the centre of public attention. Carlowitz clearly knew it well.

From England he travels on to France. These are the years when Colbert's forestry reforms reached their decisive phase. In his book *Sylvicultura oeconomica*, Carlowitz quotes extensively from the 'Ordonnances' and suggests that they contain the "whole summation" of his own work. At the start of 1668 he is in Italy, spends a long time in Rome, sails to Malta. A sojourn in Venice represents the finale of this 'grand tour'. He may have wandered through the forest of Montello on his return journey; in his book he praises it as a model.

Creating the word

In 1713, Carlowitz sums up his theoretical and practical experience in dealing with the natural resource material wood in a folio volume numbering 450 pages. *Sylvicultura oeconomica,* or 'A guide to the cultivation of native trees', appears in Leipzig. The parallels with Evelyn's *Sylva* are obvious. Carlowitz combines descriptions of useful tree varieties with practical suggestions for a long-term solution to the timber shortage, and he embellishes his account with quotations from the humanists, the Latin classics and the Bible. The arc of the text moves gradually ever closer to the idea of sustainability.

Sylvicultura oeconomica criticises the thinking of the time as being too focused on short-term gain, on making money. A cornfield will bring in a return every year, but one has to wait for decades before the wood from a forest is ready to be harvested. However, the relentless transformation of wooded areas into fields and meadows which is driven by this thinking is a mistake. The common people do not thereby spare the growing trees from the axe, because they know that they will never benefit from the timber in their own lifetimes. So they "use wood wastefully, believing it to be inexhaustible". In this way "considerable money can be had" from the sale of wood, "Yet when the woods are ruined, so the income thereof is postponed for countless years, and the treasury is entirely exhausted, so that an apparently acceptable profit actually masks a loss which cannot be made good."[1]

Carlowitz advocates a package of practical measures. 1. What today would be called an 'efficiency drive' by means of wood-saving skills, e.g. improvements to heat insulation in house construction and the adoption of energy-saving smelting furnaces, tiled stoves and kitchen hearths. 2. The use of what he calls *Surrogata* for wood, for example peat. 3. Above all, however, he advocates a methodical reafforestation by means of the sowing and planting of wild trees.

He sets up an iron rule against the over-exploitation of the forests: "that wood should be used with care (*pfleglich*)". The term *pfleglich*, according to Carlowitz, is "an age-old forestry term, commonly used in these parts of the country".

Pfleglich is in fact the direct predecessor of *nachhaltig*. No doubt the word is familiar to Carlowitz from the vocabulary of the *Jägermeister*, the masters of the hunt, who were regular visitors to his parents' house. No doubt he was also familiar with its use in the standard cameralist reference work of that time, *Teutsche Fürstenstaat* ('The German Princely State'). This book appeared in 1656. Its author, Veit Ludwig von Seckendorff, was then head of the *Cammer*, the Chamber or Treasury, in the Thuringian Duchy of Saxe-Gotha. Following its collapse in the Thirty Years' War, Duke Ernest the Pious (a direct predecessor of Queen Elizabeth II of England) was attempting to create a model modern state in this small, heavily wooded territory. He saw himself as fulfilling the role of a "good husband and father of the state". His programme was a *reformatio vitae*, a reform of life on the basis of Luther's catechism. In Seckendorff's princely state, "to use the forests with care" means "to manage them in such a way that they provide a continuous revenue over many years. . . . The harvest should not exceed the re-growth of timber. Instead the forests should provide steadily, year for year, presently and for always, wood for the use of the lord and a continuous supply of wood for burning and other uses for the people and for posterity."[2]

This tradition of careful use of wood provides the basis for Carlowitz's argumentation. He calls for a cautious use of wood, which is "as important as our daily bread", so that a balance between planting, growing and harvesting of trees is achieved and its benefits can be enjoyed "continuously and perpetually. For this reason we must arrange our economy suchly that we suffer from no lack of it, and where it has been removed entirely, that we apply our best efforts to ensure that new growth will replace it." Proverbial wisdom is used to underline the message: "Old clothes should not be discarded until one has new; just so, the stock of grown timber must not be felled until it is seen that enough has grown back to replace it."

But the traditional term 'careful' (*pfleglich*) no longer seems to satisfy the author. He searches, gropes for a new one. And then he asks "how such a Conservation and cultivation of wood can be arranged so as to make possible a continuous, steady and *sustaining* use ("eine continuirliche beständige und *nachhaltende* Nutzung"), as this is an indispensable necessity, without which the country cannot maintain its Being."[3]

There it is! The word used in its modern meaning for the first time. Its initial application seems fairly incidental: at first sight an inconspicuous attribute in a sequence of contiguous or overlapping terms. The reader can sense the dynamics of the search, can almost hear how the author tests the available words for one which will vividly and precisely express extensive temporal continuity in the utilisation of nature together with the idea of husbanding resources. The word has to be able to express in comprehensible German the meaning of the term *conservation*, derived from Latin and now established in other European languages, and combine it with the qualities of *perpetuity* and/or *continuity*. He chooses *nachhalten*.

This word had hitherto corresponded roughly with the Latin *reservare* – to hold something back, to reserve, to save for later. The semantic exercise now undertaken by our Saxon Virtuoso opens up new associative realms. It takes *nachhalten* into the lexical field of the epochal term *conservare* (conserve) and into the neighbourhood of *sustentare* (sustain). And it joins together the adjectival participle *nachhaltend* with *Nutzung* (use) in a construction which opens the door for the abstract noun *Nachhaltigkeit* ('sustainability', the state of being sustainable). And this is virgin territory. No other European language of the time boasts a term which expresses precisely what is signified here. Carlowitz not only invents the word; he sketches out the entire structure of the modern sustainability discourse – in baroque language, but in clear contours.

The three pillars of Sustainability

What does Carlowitz have to say about ecology? Nature is generous, bountiful, lavish. She is benign: *Mater natura*. The author speaks of the "wonderful world of plants", of the "life-giving power of the Sun", of the "wondrous and nourishing life force" in the soil. A plant is *corpus animatum*, a "living body which grows out of the earth, feeds itself, grows and multiplies". The external form of trees accords with "their inner form, the signature and constellations of the sky under which they grow green", and with the *matrix*, the soil and its natural power. Nature is "inexpressibly beautiful. How pleasing the green colour of the leaves is, cannot be said." Nature is "unfathomable": she "still keeps many things hidden from Man". But we are able to "read in the Book of Nature" and "to investigate by experiment how Nature plays" and to "follow in our thoughts her miraculous workings". Here it is again – *thaumazein*, the sense for wonder celebrated by the philosophers. Modern ecologists speak of the 'intrinsic value' of nature. In this text of the Baroque period this value is still self-evident. Nature is – also – a sacred realm.

How is the Inspectorate Director's economic thinking structured? His starting-point, as for Spinoza's *Ethics*, is the simple recognition of the fact that Man no longer inhabits the Garden of Eden. He can no longer simply leave everything to Nature. He must come to the aid of the vegetation of the Earth and work together with her. Carlowitz quotes – as did Evelyn before him – the Biblical command from the Creation story to "work and take care of the Earth" (*Genesis* 2:15). This text is often quoted today as a source of the sustainability idea. For Carlowitz, it is the foundation of all economic activity. Fire, and therefore wood, is an indispensable resource. But Man has been set limits to the utilisation of these divine gifts. Economics is an imitative science. It must not go against nature, but must follow it, and this includes the proper husbanding of resources. This means not cutting more wood than the forest can bring forth and support. Carlowitz calls for "a balance between the planting and growth and the harvesting of trees". To ignore the *matrix*, that is, the regenerative power of nature, will inevitably lead to overexploitation, to depletion. *Sylvicultura oeconomica*, or literally, Economical forestry, means the rational, wide and responsible management of the forests. The vocabulary of the doctrine of Providence reappears. The object of economics is the *sustainment* and *conservation* of a country. And this must be achieved from its own resources: "A country must not provide for its basic needs from abroad." Carlowitz also decisively rejects the strategy of subjugating foreign territories. The ethical and moral core of his political economy is "the raising up of the country and of its subjects," or "the betterment of the general welfare of the country".

In accordance with these aims, Carlowitz formulates the principles of a social ethics. Its foundation is the belief that everyone is entitled to nourishment and sustenance, including the "poor subjects" and "beloved posterity", that is, future generations. Notwithstanding his proper concern to replenish the coffers of his patron, the wellbeing of the commonwealth, the 'community' is a higher priority for Carlowitz. What is at issue here is the maximum degree of happiness attainable in this life, which is strictly separated from that eternal bliss which is only to be found in the hereafter. In this context, the idea of acknowledging our responsibility towards future generations is crucial. It is one which Carlowitz returns to, and varies, repeatedly, and which forms the leitmotif of his thought.

In this ancient book, then, we find in outline the entire concept of sustainable development: respect for nature – management of resources – strengthening the community; and in addition, taking responsibility for succeeding generations. None of this is an original innovation on the Director's part. Carlowitz's thinking follows the tracks of German cameralism, that is, of the contemporary economic and social doctrine espoused in all the German territories. But in him this teaching rests on a particular foundation. Between the lines, Spinoza's framework of sustainability can be clearly discerned. Certainly it is a stronger presence than Descartes' guiding principle of the subjugation of nature through human reason. In describing the fertility of Mother Nature, Carlowitz comes quite close to Spinoza's *natura naturans*. When he says that "Nature, or rather God the almighty and omniscient Creator" is the maker of all of the rich variety of living things, one can hear echoes of *Deus sive natura*. The sentence at the heart of the book is particularly striking. The conservation of wood should follow from its sustaining or sustained use *('nachhaltende Nutzung')*, because otherwise it will not be possible for the country to maintain its Being *('in seinem Esse ... bleiben')*. As we know, *suum esse conservare* is the idea at the core of Spinoza's *Ethics*. *Conservatio* was the key term of the doctrine of Providence. Carlowitz brings all the elements of the conceptual structure of his precursors into play. This is the strength of his creative act of linguistic innovation.

The word *nachhalten* appears in the book only once more. In the chapter entitled 'Burning charcoal', Carlowitz tells of the gypsies (*Zigäuner*) in Egypt and Hungary "who carry out the blacksmith's work in the open fields" and who "have a goodly knowledge of how to burn good coal which will sustain its heat (*so lange nach hält*) and will keep burning longer than other coals. They know well too how to harden iron ..." A small, touching tribute from one virtuoso to others. The Saxon nobleman seems not to have suffered from a fear of contact with other cultures.

A glance towards the Far East

Is sustainability then a 'European Dream'? Yes; but not by any means a European monopoly. In the 1660s, when Evelyn's *Sylva* was published, while Colbert was planning his forestry reforms and Carlowitz was travelling around Europe, on the other side of the planet, in a Japan still hermetically sealed off from the outside world, a code of conduct was being formulated on the right way to live one's life. The author is Tsugaru Nobumasa, a powerful daimyō (a feudal lord) in the far north of the main island, a lover of swordfighting and of lacquerware. "One must take care", he says, "of the family line and for one's heir." He places the fundamental rule of Confucianism right at the start. And in the next sentence already he speaks of the right way of dealing with nature and with the five elements – fire, water, earth, metal and wood.

> One's third consideration is the mountains. To elaborate, man is sustained by the five elements. In our world today neither high nor low can survive for a moment if any one of the five is missing. Among the five, water and fire are most important. Of the two fire is more crucial. However, fire cannot sustain itself; it requires wood. Hence, wood is central to a person's hearth and home. And wood comes from the mountains. Wood is fundamental to the hearth; the hearth is central to the person. Whether one be high or low, when one lacks wood, one lacks fire and cannot exist. One must take care that wood be abundant. To ensure that wood not become scarce, one cherishes the mountains. And thus, because they are the foundations of the hearth which nurtures the lives of all people the mountains are to be treasured.[4]

Another *Ur*-text of sustainability – this time from the Japan of wandering monks, samurai and haiku poets. Tsugaru Nobumasa was born in 1646, one year later than Carlowitz, and he died in 1710, four years earlier. Each knew very little of the other's world. Nevertheless, Carlowitz noted with surprise that wood was considered the fifth element in China. There are striking similarities between the fundamental issues identified, the solutions proposed and the language in which the proposals are made. The translation of noble aims into reality, however, turned out in every culture of the world to be an immense challenge.

The testing-ground of Weimar

In the first half of the 18th century, *Sylvicultura oeconomica* was compulsory reading for the cameralists of the German petty states and beyond. The book was read in the 'Economic Society' of the Swiss canton of Berne and in Lutheran parsonages in the Finnish province of Ostrobothnia. A second edition appeared in 1732. Carlowitz's 'sustainable use' gradually established itself as a clearly defined term. In 1757, the German forester Wilhelm Gottfried Moser began his standard work on the *Economic Principles of Forestry* with the following sentence: "The science of forestry teaches us how to manage forests so that they continue to provide sustainably (*nachhaltig*) the greatest possible benefits."

Three years later, in 1760, the duchy of Saxe-Weimar became the first testing-ground for a form of forestry guided by this new concept. The location is not coincidental. The demand for wood for glass-making, the frequent wars and the incurable profligacy of the ruling dynasty had devastated the forests, the main resource of the region of Thuringia. In 1758, following the death of the Duke, his young widow Anna Amalia became regent in place of her newborn first son Carl August. She was a member of the House of Welf, or Guelph, born Duchess of Brunswick and Lüneburg. Her mother was the sister of the Prussian King Frederick the Great. Her hopes of presiding over a court befitting her lineage and station were out of the question here. But her dream of a 'Court of the Muses' was based on a vision of making a creative virtue out of the scarcity of resources at her disposal, of generating a qualitative maximum from a quantitative minimum. It was during her regency that the forestry administration began the transformation towards sustainability.

A key scene was played out in the castle in Weimar in 1760, in the middle of the Seven Years' War. In order to fill the chronically inadequate state coffers, the ducal Treasury, the 'Cammer', yet again demands that the revenues from the forests should be increased, this time by 2,000 Talers per year – not in all truth a great sum. But then something unexpected happens. The demand is met with resistance. The State Hunting Master and head of the Weimar Forestry Department, Johann Ernst Wilhelm von Staff, declares that this is an overestimation of "the true capacities of the forests". Any further increase in tree-felling would result in depletion. He presses for a thorough review of the condition of the forests, to be carried out by a technically qualified committee. The Treasury gives way, and approves the proposed 'Taxation', or valuation. Thereupon the 23-year-old Anna Amalia signs a decree authorising a comprehensive stocktaking of the ducal forests. They are to be "geometrically measured out, with detailed records made of type and quality of timber, and to be subject to a new and sustainable forest management plan established according to the proper principles of forestry."

With her signature Anna Amalia introduced the first-ever comprehensive or state-wide forestry reform explicitly based on the principle of sustainability. It has now become a principle of governance, an element of the national interest.

The group of experts appointed by the Duchess sets to work. Like drivers for a hunt they comb the forests of Thuringia; not to drive up game, but hunting out data on the stocks of timber and on the clearings. After three years, the group presents its findings in the form of tables and 'sketches', or maps. These describe the individual wooded areas and calculate annual timber harvests on the basis of predicted growth rates. The principal aim is to establish a "constant, balanced use of the forests" and thereby a sustainable yield.

What does it mean to make "the true capacities of the forests" the benchmark for their use? No longer is the demand for wood or for money the decisive factor. Rather, the rate of regrowing timber at any given time marks the upper limit for the annual harvest. And that means that human intervention has to align itself to natural cycles. The use of wood, that is, the felling of trees, is thereby connected directly and inextricably to the regeneration of the forest. The replacement of a resource becomes a precondition of its use. Economics is consciously and deliberately embedded in the natural world. It is understood, in the words of Johann Wolfgang von Goethe, Minister in the government of Weimar, as "taking part in the productive processes of nature".

In Goethe's forests

'Mist rising from valleys near Ilmenau' is one of the most beautiful of Goethe's drawings. Sketched in pencil and ink for Charlotte von Stein on 22 July 1776, it shows the view from the Hermann Rock, a notable feature on the north-west slope of the Kickelhahn mountain, looking towards the ridge of the Thuringian Forest. Sitting "high on a mountain with a wide view", Goethe writes a letter to his beloved friend just before he begins to draw.

> "I'm sitting in the rain under the shelter of fir branches waiting for the Duke, who will bring along a rifle for me. The valleys round about are steaming upwards along the walls of spruce."

In the drawing, the outlines of the present-day landscape to the west of the Thuringian town of Ilmenau are easy to make out. However, there is no undulating sea of trees to be seen in the picture. Clearly, the slopes of the Kickelhahn below the rock were at that time bare. A few solitary pines are all that stand out in the foreground. Only further down the slope and on the neighbouring hilltops and ridges does the tree cover thicken into tall pine forest. Early morning mist, perhaps mixed with swirling smoke from charcoal kilns, rises from some of the valleys. But despite this one can see clearly that the furthest lines of hills, just before the ridge, are only thinly wooded, at best dotted with stumps or with young trees and plantation areas. Only narrow bands of mature spruce trees run over the slopes. Even on the flanks of the Finsterberg mountain in the background, the chequerboard pattern of clearings and new plantations is unmistakable. Goethe drew the forests of Ilmenau at exactly the moment when the new reafforestation programme, driven by the vision of sustainability, had just begun.

About 20 years after Goethe captured the forests on paper, another great German poet and playwright, Friedrich Schiller, visited the Ilmenau *Forstmeister* Carl Christoph Oettelt and watched him at work. Oettelt's pupil, Gottlob König (later to become famous in his own right as the author of a standard work on forestry), describes the scene for us. Oettelt is in the act of putting down on paper the tables and maps for one specific forest management project. He is planning the felling and planting of trees for the woods surrounding the Kickelhahn. The logging areas, or compartments, have been decided on in advance for two rotation periods, each of 120 years. Each compartment is marked with the year for the planned felling. So the Weimar forester is mapping out, around the year 1800, the forests of the year 2025. "By God," Schiller is supposed to have cried out, "I thought you foresters were brutes, concerned with nothing more elevated than the killing of wild animals. But you are noble indeed. You work unrecognised and unrewarded, free from the tyranny of egotism, and the fruits of your silent labours ripen for a distant posterity." What captured Schiller's imagination was the capacity to foresee and to forestall.

A site inspection, summer 2009. Between the ridge of the Thuringian Forest and Ilmenau, the hiking paths lead you from time to time past extensive bare stretches of land. Entire hilltops and slopes have been denuded of trees. In January 2007, hurricane Kyrill swept over the ridge. Next morning, huge numbers of trees lay on the ground. Forestry experts today warn that the German spruce forests may collapse entirely over the next two or three decades due to the impact of climate change. It does not look as though the forest which Oettelt had imagined, planned and calculated for 2025 will become reality. But in fact the weaknesses of 'rational' forest planning were recognised very early on. The search for solutions was already under way in Oettelt's own lifetime. The rise of ecology, within the framework provided by the concept of sustainability, began in the middle of the 18th century.

CHAPTER EIGHT

The birth of ecology

Linnaeus's "oeconomia naturae"[1]

Around the year 1730, Linnaeus began to classify plants according to the structure of their sexual organs. Carl Nilsson Linnaeus, born in 1707, was the descendant of a long line of Lutheran ministers in the southern Swedish province of Småland. His big idea, viewed as scurrilous at the time, remains a source of contention, amusement and outrage today. The American humorist Bill Bryson recently made fun of the Swedish naturalist's "abiding and feverish preoccupation with sex". There may be some truth in that comment. But it misses what was at the core of his work. My reading is that, whether bent over the corolla of a flower in the botanical gardens of the University of Uppsala, of which he was the Director, counting, dissecting or measuring stamens and pistils, or on the rocky cliffs of the Baltic island of Gotland, or the treeless tundra of Lappland, Linnaeus was always searching for the same thing: the key to the flourishing of his country. His sheer delight in the multiplicity and diversity of species and natural forms, his passion for the processes of fertilisation, reproduction, growth and breeding, for the whole *multiplicatio individuorum* – the 'reproduction of individuals' (Linnaeus, 1735) arose out of a defining interest in the *sustainable use* (Carlowitz, 1713) of living resources.

"Whoever seeks entry into the realms of Nature must first pass through the antechamber of language", he declared towards the end of his life in a speech in the cathedral at Uppsala. Linnaeus's system of nomenclature, organisation and classification, brilliantly simple and still in universal and unchallenged use today, was designed to accommodate a comprehensive inventory of flora and fauna. The aim of the binomial system – first the genus, then the species, in botanists' Latin – was to enable shared international recognition and understanding of all the blue planet's green treasures in a language that was simple to use and accessible to all.

There can be no doubt that Linnaeus saw his work, although very much an Enlightenment project, as being in the service of God. In the reproductive processes of an autonomous nature he was searching for *skaparens egna fotspor* – "the very footprints of the Creator". Even the pedantically ordered botanical collections and gardens of the time expressed this belief. But it was being progressively secularised. The conviction that God was actively present, *sustaining and creating*, in the tiniest mite just as in the vast solar system, was for Linnaeus in itself an item of rational faith, or of faith in Reason. His system, anticipating Darwin, already placed man, the *animal rationale*, in a single category together with the apes. This was a step too far for many of his contemporaries. At his university, which was a stronghold of orthodox Lutheranism and the philosophy of Leibniz and Wolff, the objection was raised that he "conflated Nature and God". This was an accusation which had to be taken seriously. It implied a charge of pantheism, and this in turn was considered a variant of atheism. Theologians at the university of Halle had used a similar wording in their attempt a few decades earlier to silence Tschirnhaus and other followers of Spinoza.

Linnaeus's lifelong motto was *oeconomia naturae*. He probably took this Latin phrase from the English physicist and theologian Thomas Burnet. Burnet, a contemporary of John Evelyn's, wrote in 1697 about "this Oeconomy of nature, as I may call it, or well ordering of the great Familiy of living Creatures. . . ". The Latin word *oeconomia* is derived from the Greek οἶκος, meaning house or household. In the context of nature, it signifies the unity and totality of nature, the diversity of species, the cycles of growth and decline, symbiosis, food chains, energy streams, the succession of plant communities and their capacity to regenerate – in short: the life of nature in all its richness and fullness. The mineral world, the vegetable world and the animal world form an interconnected whole. They are one self-regulating and self-sustaining organism, and this organism ensures that life itself is sustained and able to continue. Linnaeus observed very carefully the "war of all against all" in the world of nature. He describes how animals "not only destroy the most beautiful plants, but tear each other to pieces without mercy". However, the food chains, the eating and being eaten, always serve to maintain a well-ordered whole. Self-government and self-preservation, growth and decay, the dynamic interplay of creative and destructive forces, all of it runs smoothly and ceaselessly – and without human intervention.

> By the Oeconomy of Nature we understand the all-wise disposition of the Creator
> in relation to natural things, by which they are fitted to produce general ends, and
> reciprocal uses.[2]

This is how Linnaeus and his disciple Isaac J. Biberg defined his central concept in 1749. The idea of sustainability is clearly encapsulated. Linnaeus is interested in the phenomena of preservation and endurance. He focuses his attention on symbiotic relationships and on fertility in nature. "In order, therefore, to perpetuate the established course of nature in a continual series, the divine wisdom has thought fit, that all living creatures should constantly be employed in producing individuals; that all natural things should contribute and lend a helping hand to preserve every species; and lastly, that the death and destruction of one thing should always be subservient to the restitution of another."[3]

But how is the *oeconomia naturae* related to what Linnaeus calls *oeconomia nostra*, the economy of Man? Linnaeus had a clear view on the relationship between economy and ecology – the all-important issue of our own century: the synchronisation of our economy with the great, unchanging, God-given cycles of the *oeconomia naturae* has to be achieved. "Nature does not allow anyone to master her," he writes in emphatic opposition to Descartes.[4] "Nature's economy shall be the base for our own, for it is immutable, but ours is secondary."[5]

The dominant economic philosophy in 18th-century Sweden was that of cameralism. When Linnaeus began his career, the Kingdom of Sweden was on its knees. For a hundred years the country had waged an unbroken series of expansionary wars. The Swedish kings and their *soldatesca*, or mercenaries, had practised a naked economics of robbery in the 17th century, and indeed had carried it to extremes, largely on German soil. Their method was extortion, and the threat behind it was torture and arson. And while Swedish and Finnish peasant soldiers, at the cost of enormous losses to their own ranks, were off fighting and laying waste to central Europe, back home the corn was rotting in the fields. With the defeat of Sweden in the Great Northern War in 1721, this era came to a decisive end. Suddenly, the Swedes found themselves forced to withdraw to their own territory and restricted to their own resources. In order to adapt to the new situation they imported the doctrine of cameralism from the German princely states complete with its ideological structure of self-sufficiency and self-government, careful use of one's own resources, and sustainability. The *flourishing* of the country was based on the three kingdoms of nature: the animal, the vegetable and the mineral. *Oeconomia* was translated as *hushållning* (to manage the household). Its purpose was seen as the harvesting of the regular gifts of nature and their refinement for human use. The idea of economic growth was as yet completely alien to this way of thinking.

Linnaeus adhered to this cameralist view of economics all his life. When the German botanist Johann Georg Gmelin was returning to St Petersburg from his expedition to Siberia in the service of the Tsar, Linnaeus conspired with him to acquire seedcorn from the trip. He was especially interested in the fertile Siberian buckwheat, hardy and accustomed to the cold, and in the Asiatic larch, the wood of which makes excellent rifle butts. When Sweden was struck by famine in 1756, Linnaeus produced a list of edible native flora in order to ensure some alternative nutrition at times of harvest failure. "The Lapps are our teachers," he wrote, pointing to the nomadic subsistence economy of the Sami in the far north of the country, based on reindeer herding, as a model of the good life – simple, healthy and close to nature.

Even in his old age, Linnaeus could be seen in the streets of Uppsala coming back from his botanical wanderings, "an ageing man, not tall, in dusty shoes and socks, with a long beard and wearing an old green jacket with a medal hanging from it". The self-styled *princeps botanicorum,* or prince of botanists, was held in the highest regard at court in Stockholm. His great patron was Lovisa Ulrika, the Queen, a clever woman with an interest in science and an iron will, the sister of the Prussian King Frederick the Great – and an aunt of Anna Amalia, Duchess of Weimar.

Linnaeus died in the winter of 1778. Over his lifetime he examined and classified around 13,000 species of animal and plant. Each one had its own particular value. But then each one had, some 6,000 years earlier, been a part of the inventory of the garden of Eden. On Noah's ark, God had saved one pair of each of these species from the flood. This spiritual foundation of "reverence for life" (Albert Schweitzer) was still almost universal. But the suspicion was already spreading that life on Earth was the result of a long evolution. Linnaeus's successors today put the time of the beginnng of life at around four billion years ago. The American biologist Edward O. Wilson estimates the number of species at 1.8 million. If microbes are added, the number would easily exceed 100 million. But reverence for life and its diversity has not necessarily increased because of our greater knowledge.

Linnaeus's concept of *oeconomia naturae* proved equal to the task set by the explosion of knowledge about evolution which was to follow. Barely a hundred years after his death, his guiding concept provided the blueprint for a new one: ecology. The connecting link was the translation into German of *oeconomia naturae,* which was *Haushaltung der Natur,* or 'Economy of Nature'. Nowhere in Europe was this idea debated so intensively, or in such a multi-layered and interdisciplinary way, as in Weimar.

The cosmos of Weimar

Nature! We are surrounded and embraced by her – powerless to leave her and powerless to enter her more deeply. Unasked and without warning she sweeps us away in the round of her dance and dances on until we fall exhausted from her arms. . . . There is everlasting life, growth, and movement in her and yet she does not stir from her place. She transforms herself constantly and there is never a moment's pause in her. . . . She is firm. Her tread is measured, her exceptions rare, her laws immutable. . . Her creatures are flung up out of nothingness with no hint of where they come from or where they are going – they are only to run; she knows the course. She has few mainsprings to drive her, but these never wind down; they are always at work, always varied. Her drama is ever new because she creates ever new spectators. Life is her most beautiful invention and death her scheme for having much life.[6]

This short green manifesto appeared anonymously in 1782, four years after Linnaeus's death, in the *Tiefurt Journal*. Tiefurt was Anna Amalia's summer residence. The little castle sits in hilly countryside near the river Ilm, within the town precincts of Weimar. The journal circulated at the 'court of the Muses', copied in very small numbers by hand. The text outlines a whole philosophy of nature in two pages of breathless and infectious prose. The author is generally taken to be the 24-year-old Swiss theologian Georg Christoph Tobler, who spent the summer of 1781 in Weimar and conducted long conversations with Goethe. Was this text, which Goethe later incorporated in the complete edition of his works, a record of the essence of those conversations? Without doubt, it expresses the mental universe of the circle around Goethe and Herder at the time. In the heyday of the 'rational' Enlightenment, *mater natura* was suddenly once more on the stage in Weimar, powerful and beautiful. She is a subject, not an object, a living organism and not dead matter. Spinoza's *natura naturans* is as much present as Linnaeus's *oeconomia naturae*.

We obey her laws even in resisting them; we work with her even in working against her. . . . She is whole and yet always unfinished. As she does now she may do forever.

Goethe later revealed the sources of his inspiration. "Recently," he wrote to his friend Zelter in 1816, "I have been reading Linnaeus again, and am astonished by this exceptional man. I have learned an infinite amount from him, with the exception of botany. Apart from Shakespeare and Spinoza, I know of no-one among those no longer living who has had such an influence on me."

In the 1780s, the theories of Spinoza and Linnaeus became fashionable in Weimar. In the parks and salons around the ducal palace, in the garden or at the hearth, suddenly everyone is reading, discussing, arguing about and meditating on botany, zoology, mineralogy and meteorology. They prowl the woods, the rocks and the caves of the Thuringian Forest between Ilmenau and Eisenach, searching *in herbis et lapidibus* – among the plants and stones – for the divine: *Deus sive natura*. Is this just a way of passing the time? A form of therapy? A retreat from society, as the academic Goethe research community seems to think? A key to the answer may be found in a comment which Goethe made in his old age about his early career. "I knew very little about the natural sciences when I first arrived in Weimar, and it was only the desire to be able to offer the Duke practical advice in his manifold undertakings, building projects and schemes which drove me to the study of nature." Of course, Goethe the poet and thinker wanted to know "was die Welt im Innersten zusammenhält?" – "what binds creation's inmost energies?"

But in the immediate given moment he was first and foremost a Minister and chamberlain of a tiny, resource-poor, famine-stricken – today we might say 'economically underdeveloped' – territory in the middle of Germany. His studies in the natural sciences served – as they did for Linnaeus – a political end. This was what the cameralists called the 'raising up of the country', that is, its development on the basis of its own resources. It was here that Goethe wished to become "active" – and this was a matter of vital importance for him. He was granted a first insight into his new task during his first winter in Weimar in the course of long fireside discussions about *die nötige Holzkultur* – what needed to be done about the forests. Looking back later in life, he gave a precise assessment of the status of Anna Amalia's sustainable forest reform: "A review of all wooded areas, based on proper surveys, had already been carried out, and annual quotas fixed for felling in each of them over a number of years."

My thesis is that Goethe had intuitively identified the blind spot in the sustainability theory of the time, namely its domination by the Cartesian belief in the right to subjugate nature. His project therefore was to establish, using Spinoza and Linnaeus for intellectual support, the 'economy of nature' as a foundation for *all* economic activity – to plant ecology at the heart of sustainability.

" . . . a star among stars . . ."

"Our Earth is a star among stars . . ." This is the opening of one of the key works of Weimar Classicism, Johann Gottfried Herder's *Outlines of a Philosophy of the History of Mankind*.[7] The origins of this work lie in an intense exchange with Goethe in 1782/83, who at that time was working on the concept for a 'Novel about the Universe' which was later abandoned.

> Our Earth is a star among stars. If our philosophy of the history of Man would in any measure deserve that name, it must begin from Heavens. For as our place of abode, the Earth, is of itself nothing, but derives its figure and constitution, its faculty of forming organised beings, and preserving them when formed, from those heavenly powers, that pervade the whole universe; we must first consider it not singly by itself, but as a member of that system of worlds, in which it is placed.[7]

Herder here imagines the view from outside, from a flight through the solar system. He considers the universe in its entirety and speaks of the "system of worlds", the planetary system, the "great structure of the universe". This is no longer based on theology. Herder invokes the founders of modern scientific cosmology – Copernicus, Kepler, Huygens and Kant. From this distant, external and therefore sublime and elevated standpoint, Herder turns his gaze on the earth. It is bound to and by the forces of the universe. The Sun is the source of light and life in our creation. His model of the universe is heliocentric: we live by virtue of the Sun, of solar energy.

He now considers "the all-encompassing Earth below me, spinning on its axis", and his gaze penetrates the blue envelope, the atmosphere, "this great repository of active powers". He writes: "It needs not to be demonstrated, that the influence of the atmosphere cooperates in the most spiritual determinations of all the creatures upon Earth: with the Sun it shares the government of this globe, which it formerly created."[8] Herder refers often to the "climate, which determines our existence". By climate he means more than weather and meteorology. Rather, the term indicates the connection between the atmosphere and the biosphere, that is, the totality of the natural living conditions on "our dwelling, the Earth", thus anticipating the modern term *habitat*. Herder designates the limits to our action: ". . . for nature is everywhere a living whole and must be gently followed and improved, and not dominated by force." He knows that Mother Nature – the old feminine *topos* again – operates according to laws which we must not arbitrarily change.

From this Herder deduces the value of multiplicity and diversity. "Our Earth is a great manufactory, for the organisation of very different beings." The praise of biodiversity is a thread running through his entire oeuvre. In the *Outlines*, he expresses a fundamental insight of ecology. "Seldom has man exterminated any species of plant and animal from a country, without soon perceiving the most palpable detriment to its habitableness."[9] This astonishing sentence takes one straight to the most basic principle of sustainability: the preservation of the *habitableness* of the Earth as a whole. The word *habitable* no longer refers simply to the struggle with extreme climatic conditions and hostile natural forces. For Herder, *habitableness* depends on humanity's careful and considerate interaction with nature. In this he is absolutely modern. It is no coincidence that the worldwide quest which ultimately gave birth to our concept of sustainable development began in 1972 with the vision of a durable and habitable planet.

But where does mankind figure in these cosmic and planetary interconnections? Herder's thinking brings nature and humanity, natural and human history, together, and in the process comes very close to the idea of evolution. "My eye is framed to support the beams of the Sun at this distance, and no other; my ear for this atmosphere; my body for a globe of this density; all my senses, from, and for, the organisation of this Earth."[10] But this deep dependency is no reason for self-doubt. Our intellectual capacity enables us to read the book of nature and to live in harmony with her. In this context Herder draws attention to our upright gait, which distinguishes us from our "older brethren", the beasts. For him, there is a community of living beings, but there is also a hierarchy, within which mankind's proper position is in the middle. Our upright gait liberates us and gives us the capacity to look upwards, towards the sphere of the divine, the intellectual and the spiritual. And also look forwards, into the distance, into the future. "Man can investigate, explore. He can choose . . . He can take command over his own destiny."

Our upright gait – and the power of speech – enable us to be free. Free, above all, to shape ourselves. For good as well as for ill. Herder, a champion of the Enlightenment, speaks of the perfectibility of mankind. But he knows too about human corruptibility. From the idea of the unity of nature, of which we are a part, there follows for him the unity of the human race. Herder rejects colonialism, the contemporary form of globalisation, and the presumption that the inhabitants of all regions of the Earth had to be Europeans in order to be able to live happy lives. Just as he takes pleasure from biodiversity, so he also celebrates the many different cultures of the world. While recognising that there are a great many shades of skin colour, Herder decisively rejects the notion of race. This would only introduce false distinctions and deny the underlying unity of humankind. A native from Tierra del Fuego belongs to the same human race as Newton or Voltaire, he says, anticipating the Universal Declaration of Human Rights.

Herder's perspective leads from the cosmos via nature to Man. But the direction can also be reversed. There is probably no other European philosopher and writer of the time – not even Rousseau – who thought as deeply, following Linnaeus's lead, about the 'economy of nature'. As a theologian, Herder was aware of the biblical usage of the Greek word οἰκος, to mean house of God. Man is only *oikonomos*, a householder, leaseholder or steward, and not *maître et possesseur* as in Descartes. Sustainability is founded on faith in the Earth's capacity to organise and sustain all creatures – in perpetuity. Herder calls for human economy and civilisation to be embedded in the dynamics of natural processes, not simply for now but in a perspective which spans generations. "Nature plans and cares for the whole, and transforms the parts, she unfurls the chain of the generations and lets the individuals drop away; only in this way does her ever-changing and ever-enduring economy come to pass. . . . The narrow economy of men can do no other but to follow this omnipotent law of nature." Sustainability and ecology are inseparable.

Herder remained faithful to the ideas of the Enlightenment all his life. He began his philosophical journey as a student and protegé of Kant's in Königsberg. As a young philosopher and teacher in Riga inspired by the *Sturm und Drang* movement he had dreamed of becoming the Great Reformer of Livonia and of transforming the Russian province into a model Enlightenment state. But Herder saw a crucial failing in the Enlightenment which he wanted to correct: "Enlightenment is always a means, never an end. . . ." A means, that is, for the increase of human happiness. Against the "cold, empty, icy heaven" of Kant's metaphysical speculation he sets the world of sense and sensibility, the rich variety of speech and experience. To the Cartesian *Cogito ergo sum* he responds with the cry: "I *feel* myself! I am!" This does not entail a devaluation of reason, but on the contrary its extension into the dimension of sensual perception and intuitive knowledge of the exterior world – of the environment.

Songs of the Earth

"Poets establish what endures." Like a message in a bottle, the green philosophy of Weimar Classicism is preserved in two of the most famous German poems. Both have transcended the German language and borders to become a part of our common cultural heritage. One approach to understanding them takes us back once more to the Kickelhahn, the mountain near Ilmenau. I was last there in the spring of 2009.

From the hunting lodge at Gabelbach it is only a short walk over the plateau of the mountain to a simple wooden hut which played a very significant role in Goethe's life and work. It is a walk one should take shortly before sunset. It leads past storm-damaged spruce and beech trees. Weeds and blueberry bushes grow thickly in the clearings. In between, tree stumps and plantations of young spruce. From the hut a wide view opens up to the west. When the setting sun is only a finger's width above the ridge of the Thuringian Forest, the light is incandescent. No settlements or roads to be seen anywhere. Only forest. No man-made noises reach up here. Instead you can hear the noise of the wind and the evening chorus of the birds. Chaffinches and tits twitter. From its singing post, a songthrush strikes up its melody. Then the ball of the Sun touches the line of the ridge. Roughly at the peak of the Schneekopf mountain. The sky to the west turns golden. Now everything happens very quickly. A red line forms above the black ridge. A blueish veil of mist slides over the woods below. It grows dark. The birds have stopped singing apart from one hedge sparrow, which chirps for a moment longer. Bats whir through the air. In the undergrowth on the slope, a short snorting noise, perhaps a stag. Later, the call of an owl. A silver sickle Moon appears. A little later the first star comes out. A solemn moment: this is the view over the nocturnal landscape enjoyed by Goethe when in the late evening of 6 September 1780 he wrote *Wandrers Nachtlied II* ("Wanderer's Night Song") in pencil on the boards of the hut wall I am leaning against.

Über allen Gipfeln / Ist Ruh' / In allen Wipfeln / Spürest du / Kaum
einen Hauch. / Die Vögelein schweigen im Walde. / Warte nur!
Balde / Ruhest du auch.

Here are two well-known translations into English. The first is by the iconic poet
of 19th century America, Henry Wadsworth Longfellow. The second is by the
contemporary British poet and translator, Christopher Middleton.

O'er all the hill-tops / Is quiet now, / In all the tree-tops / Hearest thou / Hardly
a breath / The birds are asleep in the trees; / Wait, soon like these / Thou, too,
shalt rest.

Over mountains yonder, / A stillness; / Scarce any breath, you wonder, / Touches
/ The tops of all the trees. / No forest birds now sing; / A moment, waiting – then
take, you too, your ease.

The poem makes statements about 'die Ruhe' – peace, or quiet, or restfulness.
More precisely, it talks about three different levels of peace in the three realms
of nature. Its visual trajectory begins above the mountain peaks, the 'Gipfel' –
that is, in the atmosphere around the planet, the ether, the heavens. Up there,
there is peace. But not 'Ruhe' – the poet has shortened the word to 'Ruh'', an
extended closing vowel, a gently fading 'oo' sound – a perfect matching of
sound and meaning. "Here, the verse does not describe the stillness of the
evening, it has itself become that stillness", is the comment of the English literary critic Elizabeth M. Wilkinson. The gaze now moves downwards. It touches
the realm of the mountain-tops, that is, the mineral world, inorganic nature,
geology and geomorphology. From there it moves on into the realm of the
'Wipfel', the tree-tops, plants, vegetation. Gipfel – Wipfel. Only one sound has
changed. Seemingly dead matter has transformed itself into a nourishing
medium for living nature. In this realm, peace is no longer absolute. You feel, or
hear, 'hardly a breath'. Nevertheless, the peace is revoked. The word 'Hauch'
can be used for a gentle movement of the air and for a human breath, and its use
here equates the two. In the next line the language and focus reach the animal
kingdom. The little birds 'are silent' in the woods, that is, below the treetops.
'Schweigen' in German means to stop speaking temporarily, so the implication
is that the birds will take up their singing again with the dawn. 'Schweigen' is
actually the word for a temporary cessation of human communication, of dia-
logue. Birds simply stop singing. But now, the poem adddresses a 'you', or
'thou'. A human being. Who will rest 'balde' (soon). Here, Goethe is using an
old-fashioned, slightly portentous form of the common word 'bald'. The
addressee is therefore not yet at rest. They are still restless. The most restless

element of nature. Without movement and activity, they could not be aware of the peace of the cosmos, the breath of the wind, the silence of the birds. But they too yearn for peace. They wish to come to rest.

With this, the movement of the poem through the three realms of nature reaches its majestic finish. The downward diffusion of peace from above, from the cosmic sphere, finds its conclusion in Man. 'Warte nur, balde ruhest du auch.' Only wait, soon you will rest too. The strong emphasis which is given to 'auch' (too) makes it into what the literary critic Werner Kraft calls "the metaphysical equals sign" between Man, nature and the cosmos. It annuls the antithetical separation of self and nature. Cartesian dualism is resolved. And with it the anthropocentric belief in the primacy of Man within creation.

With this inclusive 'auch' (too), Man takes his place in the endless chain of beings to whom the poem promises rest, or peace. He is organically conjoined with all living – and therefore mortal – creatures. With the cosmos, the world order and its cycle of growth and decay. He achieves a state of balance, of harmony with creation. All natural things, says Goethe, stand "in a precise relationship" to each other. What kind of 'rest' awaits human beings? Is it simply the wanderer's nightly rest that is referred to here? Or the subject's inner peace, his serenity? Whose is the voice saying 'you'? Is it the inner voice of the implied subject, who is performing a soliloquy or interior monologue in the twilit landscape? Is it the dead mother, calling her child? In its tone and rhythm, the poem feels a little like a lullaby. Is it 'Mother Nature' herself speaking? Or God? Or *deus sive natura*, nature worshipped as divine? That remains the secret of the poem. It is open to the interpretation of the reader. Goethe speaks, as it were, of the end of speech. The 'Hauch' – the breath, the word, the spirit, *pneuma* – comes to rest. One could say that with the simple word 'auch' (too), with its insistent rhyme on 'Hauch', Goethe's poem, too, dies away.

> Freude, schöner Götterfunken, / Tochter aus Elysium, / Wir betreten feuertrunken / Himmlische, dein Heiligtum . . .

> Joy, beautiful spark from the Gods, daughter of Elysium; drunk with fire, we enter your sanctuary, O heavenly one . . .

Schiller's Ode To Joy has been described as a 'great cantata of the cosmos'. This is the poem of Schiller's which has had the greatest resonance, thanks to its having been set to music by Beethoven in the closing chorus of his Ninth Symphony. Since 1972 this setting has been the official anthem of the European Union. The ode was written in the autumn of 1785, five years after the 'Wanderer's Night Song', in a little house in the vineyards of the Elbe valley near Dresden. At first sight, the two poems have little in common, thematically or formally. Schiller, then 25, celebrates in it 'joy', and as the source of joy the power of mutual attraction between human beings in all its variations: friendship, conjugal love, *Sympathie* or liking, desire, collective ecstatic experiences, and, as an overarching concept, fraternity and sorority, the familial or quasi-familial solidarity which spans the globe.

"Alle Menschen werden Brüder" – all people become brothers – is the line of verse which Beethoven's setting of 1827 gives such great emphasis. Schiller brings into focus what we might call today the cohesive role of civil society in its many structural forms: intimate relationships, neighbourhoods, the bands of friendship, the social circle, what he calls the "holy circles", the secret circles of emancipatory or liberation movements, of revolutionaries. His text is an attack on the feudal class society of his time. The lines "Deine Zauber binden wieder, / Was der Mode Schwert geteilt" (your magic binds together again what was divided by the sword of custom) are aimed at social convention and political power, and the rigid division of society into classes, and of the world into nations, spheres of influence and races, which they enforce. It is *joy* alone which can provide the basis for new bonds between people. The word carries overtones of eroticism, empathy, mutual respect and solidarity.

However, the ode does not remain anthropocentrically restricted to the sphere of human relationships. It is imbued with a natural philosophy every bit as sophisticated as that of the 'Wanderer's Night Song'. The third verse declares Nature – addressed as the fruitful, life-giving *mater natura* – to be the source of all joy, and for "all creatures". The provocative lines "Wollust ward dem Wurm gegeben, / Und der Cherub steht vor Gott" (even the worm was given sensual ecstasy, and the cherub stands before God) call to mind Linnaeus's sexualisation of nature. They bring right up to date the classical idea of "the great chain of sentient beings" (Schiller), which had already been revived by Leibniz and others in the early Enlightenment. This chain stretches from lifeless matter via the simplest forms of life all the way to celestial beings.

As in so many other poems, and even in his plays, Schiller uses this point as the launchpad for an extended "cosmological conceit". The power of attraction between human beings, he suggests, corresponds to the force of gravity between physical bodies, the power of attraction proportional to their mass, worked out by Newton in 1666. An analogy is created between cosmic events and human feelings. Both, says Schiller, clearly echoing Spinoza's philosophy, are "reflections" of a "single primal force", which operates in both the extraterrestrial cosmos and in the world of human emotions, and which bears and affects dead matter as well as all of life.

The identity of the divinity remains open in the ode "To Joy". If the opening evokes the Greek myth of the "Isle of the Blessed" with its reference to the "daughter of Elysium", subsequent references invoke the "loving Father" of the Judaeo-Christian tradition, the "Unknown", the "great God", "the Gods", the "good Spirit" and finally the "heavenly judge". Whatever: "all sinners shall be forgiven, and hell will be no more". "No hell below us" – almost 200 years before John Lennon and Yoko Ono! Schiller's imagination pictures a planet "floating in infinity" (in the poem 'The Greatness of the World') on which life is borne forward and developed by symbiotic communities of all living things and by free human societies.

Competing visions

In the green imaginings of Weimar Classicism and early Romanticism, human-kind appears as a creature of nature endowed with a special power: with intel-lect. This picture of humanity is expressed with breathtaking power in a tribute of Goethe's written in 1805 to Winckelmann, the archaeologist and connoisseur of Greek statuary, and one of the founders of the revival of European Hellenism and of German Classicism.

> When Man's nature functions soundly as a whole, when he feels that the world of which he is part is a huge, beautiful, admirable and worthy whole, when this har-mony gives him pure and uninhibited delight, then the universe, if it were capable of emotion, would rejoice at having reached its goal and admire the crowning glory of its own evolution. For, what purpose would those countless suns and planets and moons serve, those stars and milky ways, comets and nebulae, those created and evolving worlds, if a happy human being did not ultimately emerge to enjoy existence? [11]

This is a truly Promethean perspective. The question, "What can the Earth do for mankind?" is augmented by the question, "What can mankind do for the Earth?" Reverence for nature and the cosmos is reconcilable with pride in human dignity and worth. The real sustainability programme of Weimar Classicism is a new synthesis of ecology and humanity, mind and nature.

A competing model was developed at around the same time in Britain, one which spread quickly through both old Europe and the New World. Goethe's contemporary, the Scottish moral philosopher Adam Smith, linked a broad view of humanity together with the concept of the free market in the year 1776 – Goethe's first year in Weimar. The overarching idea was "the greatest happiness for the greatest number". This phrase was launched into the world – also in Scot-land – by the philosopher Francis Hutcheson. As early as 1725 he had posed the question of how to measure and compare the moral quality of actions in order to be able to 'compute' one's decisions in any given situation. His answer: "that action is best, which procures the greatest happiness for the greatest numbers". This definition can be linked to the ancient philosophical school of the Stoics, who had required in their theory of the "common good" that as many people as possible should be made as happy as possible. Hutcheson's formulation already incorporates the Enlightenment obsession with quantification. The philosopher himself speaks of his 'moral algebra'. If *levels* of happiness procurable are equal, Hutcheson argues, then the choice should be determined by the *number* of people who will enjoy it. With this step, the universalism which was so important to Schiller, the inclusion of 'all people', is abandoned.

The thinkers of the Scottish Enlightenment professed their belief in utilitarianism. Their philosophical system makes the usefulness or benefit of knowledge, and the cost-benefit analysis which is its corollary, the decisive criterion. Adam Smith expands on the start made by Hutcheson. Individuals who work for their own happiness, in full freedom and in their own interests, simultaneously bring about by this means the greatest general happiness. Translated into economics: if every individual manages their capital so as to maximise the increase in value it yields, then the 'invisible hand of the market' will raise aggregate income and prosperity to the highest level possible. This is how Adam Smith couples the concept of the free market to the principle of the greatest happiness. It was in this utilitarian form that the philosophy of happiness entered into the infant American Constitution. It was Thomas Jefferson, the 33-year-old representative from the colony of Virginia, who introduced "the pursuit of happiness" into the 'Declaration of Independence' of the United States of 4 July 1776. Not happiness itself, but "the pursuit of happiness" became an inalienable human right. Is this a renunciation of any form of compulsory, prescribed happiness? Of the narrow worlds of the cameralist regimes, in which all too often a corrupt and bigoted courtly clique had determined what the happiness of the subjects should consist in? Yes, of course, and rightly so. But from today's perspective it appears that this liberal principle also contained within itself grave risks. Liberalism had an inbuilt tendency to turn the 'invisible hand of the market' into a fetish, and nature into an infinite resource depot. In this belief system, 'sustainability' was considered an outdated doctrine of petty central European states, and ecology was merely a tool to help cultivate useful plants and animals and to exploit them commercially. The 'greatest happiness for the greatest number' was calculated from the beginning by a monetary measure.

The first ecologist

Beginning with the depths of space and the regions of remotest nebulae, we will gradually descend through the starry zone to which our solar system belongs, to our own terrestrial spheroid, circled by air and ocean, there to direct our attention to its form, temperature, and magnetic tension, and to consider the fullness of organic life unfolding itself upon its surface beneath the vivifying influence of light.[12]

Once again, an imagined view down onto the blue planet from above. It is described by Alexander von Humboldt in his magnum opus, *Cosmos*. This 'Sketch of the Physical Description of the Universe' was published in five volumes beginning in 1845. Here, too, the cosmic perspective is intended to provide a basis for the concept of a totality of nature, the sense of integrity and harmony in the cosmos. Humboldt sees nature as "an intricate reticular web" and "an integral whole given animation and direction by inner forces". All earthly powers, from the elements, via inorganic nature, plants and animals to human beings, are "children of Gaia". The path from the economy of nature to ecology leads via Alexander von Humboldt.

In the scientific networks coalescing in the little cosmos that was Weimar, the young natural scientist and explorer from Berlin quickly became a central figure. The geologist Gottlob Abraham Werner, Goethe's idol and Novalis's teacher, was his mentor at the Mining Academy in Freiberg. Humboldt published his first writings in 1795 in Schiller's journal *Die Horen* (The Hours). Alongside Goethe he studied natural sciences at Jena in 1797 in order to prepare himself for his expedition to South America. The two remained in contact until Goethe's death in 1832. Herder's *Outlines* find their continuation in Humboldt's *Cosmos*. Humboldt adopts Herder's concept of humanity; he, too rejects racism and slavery and emphasises the sense of community and unity of the whole human race, and of the equality of rights among all its members.

Humboldt has been described as 'the first ecologist'. However, his career began – as did those of Linnaeus and Goethe – in the cameralist tradition. He was commissioned by the Prussian mining authorities to redevelop the mining industry in the mountains of the Fichtelgebirge. One of his first tasks: to overcome the wood shortage. He never forgot this lesson. Even in the tropical rainforests of Venezuela he found it incomprehensible that "the trees are being cut down as furiously as in Franconia, leading to shortages of both wood and water". (We shall return to this later.). Humboldt extended Linnaeus's classification and description of species to a *Geography of Plants and Animals* and finally to a *Geography of Organic Life*, which was an exploration of the subtle interconnections between flora and fauna, and between geographical, climatic and environmental factors. From here it was only a small step to the formation of a new scientific discipline and to the coining of a new term. This step was taken, seven years after Humboldt's death, by a young professor of biology named Ernst Haeckel. The stage was the University of the Duchy of Saxe-Weimar in Jena, the alma mater of Leibniz and Carlowitz, where Schiller and Schelling had taught one generation earlier, and through whose doors Goethe and the Humboldt brothers had passed to and fro.

Defining ecology

On a photograph from the year 1866, Ernst Haeckel looks less like a distinguished full professor and more like a Greenpeace activist at the launch of the 'Rainbow Warrior'. The picture shows the 33-year-old naturalist in light-coloured outdoor gear. He wears a neckerchief. On his legs are a pair of what look like cowboy boots. His hair is long and wavy, his beard blond and neatly trimmed. His gaze goes past the camera, presumably out of the window of the room towards the open sea. The portrait was taken on Lanzarote, the volcanic island off the Moroccan coast, where Haeckel and his assistant, a somewhat scruffy but equally intrepid-looking Russian medical student, were collecting sponges, plankton, protoplasm and a multitude of other small marine life forms and examining them under their Zeiss microscopes. Shortly beforehand, Haeckel had invented the term 'ecology'.

A zoologist from Jena, born in Potsdam in 1834, Haeckel was familiar with the green ideas of Goethe, Herder and Humboldt. Despite his relative youth he already had gained an international scientific reputation. From the waters around the Mediterranean island of Capri he had fished up more and more new and fascinating species, sub-species and intermediate species forms of radiolaria. These single-celled organisms revealed under the microscope an astounding world of forms and structures. While he was analysing his findings, around 1860, he chanced upon a book recently translated from the original English – *On the Origin of Species by Means of Natural Selection, or the Preservation of Favoured Races in the Struggle for Life*. Its author was Charles Darwin. The book induced in Haeckel an incredible euphoria. Here he found the key and the confirmation for his own research findings. Inspired by Linnaeus's *oeconomia naturae*, Darwin had turned his sharp gaze onto the whole of natural life and had devised a new theory of evolution: all species evolved from respectively simpler forms. In this process, multiple and minute changes in an organism form the initial phase for the emergence of a new species. Those changes are the result of processes of adaptation to the specific conditions of the organism's habitat. The selection of the best-adapted forms is the mainspring of evolution. This new and revolutionary theory gave Haeckel a coherent explanation for the variety of forms of his radiolaria, and enabled him moreover to use his own research as empirical evidence in support of Darwin's theory. Haeckel became the great communicator of the theory of evolution, held in high regard by the master himself.

However, Haeckel's *magnum opus* was an independent work of his own. It appeared in October 1866 in Berlin under the title *Generelle Morphologie der Organismen* (A General Morphology of Organisms).

Each chapter was preceded by a quotation from Goethe's natural science writings. Right at the start, the reader lights upon a passage from the 'Fragment' on nature inspired by Goethe: "There is everlasting life, growth, and movement in her and yet she does not stir from her place. She transforms herself constantly and there is never a moment's pause in her."

What is Haeckel's original and distinctive contribution? From the "Conservation or Self-preservation" of organisms, that is, their nutrition and reproduction, Haeckel now turns his attention to the individual parts of the organism and their relation to each other and to the whole, and to the place which each organism has in the household of Nature and in the economy of the natural system as a whole. It is in the context of this Linnaean vocabulary that the word 'ecology' first appears.

> By ecology, we mean the whole science of the relations of the organism to the environment, including, in the broad sense, all the 'conditions of existence'. These are partly organic, partly inorganic in nature, both, as we have shown, are of the greatest significance for the form of organisms, for they force them to become adapted. . . . the entire relations of the organism to all other organisms with which it comes into contact, and of which most contribute either to its advantage or to its harm.[13]

Here can be found in one short text the green vocabulary of all epochs. The *Conservatio* of the doctrine of Providence is the starting-point. *Organism*, a term which is the antithesis to all mechanistic interpretations of life, and *habitat* are also there. Climate appears as a condition of existence, i.e. as a necessary basis for life. *Evolution* means more now than the unfolding of a bud or the growth of a seed of corn. Darwin's theory has raised *evolution* to the fundamental law of the planet.

"Ecology, or the science of nature's household" – this was Haeckel's concise definition of the term he had coined – became the name for a new branch of science. Its subject was the interrelationships amongst living things and between them and their environment. Haeckel's word was adopted quickly into other languages: for example in 1893 it can be found in a British medical journal. But for a long time this scientific discipline remained a small backwater of biology. By the middle of the 20th century there were at most a couple of hundred specialists in the world who had an idea what the term meant. For almost a hundred years, 'ecology' remained locked up in the ivory tower of academic terminology.

But why? By the middle of the 19th century at the latest, in the countries of the Western Enlightenment, all attention was diverted away from the living, growing, endlessly self-renewing resources of the Earth's surface to the subterranean forests, the repositories of fossil fuels. The lustre of *oeconomia naturae* faded. The conviction that the free market and infinitely continuous worldwide economic growth would deliver the greatest happiness for the greatest number became for a long time – right up into our 21st century – hegemonic. It was based on the belief that fossil fuels could be exploited indefinitely.

Environment and development

The Earth Summit in Rio in 1992, which established the term 'sustainable develop-ment' throughout the world, was officially called 'the United Nations Conference on Environment and Development'. It is quite breathtaking to follow how not only *ecology* and *sustainability*, but also *environment* and *development*, those other two key words at Rio, were taken up and rejuvenated in the communicative networks of Weimar Classicism 200 years earlier, and how each of them acquired a highly nuanced character which carried over into other European languages.

Herder and Goethe had a particular affection for the term *development*. Both had a feeling for the sensuousness of the word. They valued its capacity to express the idea of continuity in change, and to do so flexibly, accentuating either the connec-tion between any given state and an earlier one or its specific new qualities. They were familiar with the long usage of the word and of its equivalents *evolutio, evolu-tion* and *développement* in the language of natural history and biology. "Our daily conversation", Goethe wrote, recollecting his joint studies of nature with Herder in the 1780s, "was concerned with the primal origins of the water-covered Earth and the living creatures which have evolved on it from time immemorial. Again and again we discussed the primal origin and its ceaseless development." One topic of these conversations was the distinction between humans and apes, which Darwin took as the starting-point of a theory of evolution several decades later. In the microcosm that was Weimar, however, the concept of development took off in a new direction. It was extended into the political-social sphere. Herder posed the question whether there was "a continuous thread in the development of human capabilities through the centuries". In his 'Letters for the Advancement of Human-ity' in 1793 he pleaded that humanity should become aware of what it was work-ing towards and of what was developing before its eyes. His plea reads like a poetic version of the Brundtland formulation: "The present is pregnant with the future, the fate of those who come after us rests in our hands, we inherited the thread, we weave it and spin it on."

Marx and Engels, both assiduous readers of Weimar Classicism, had nour-ished themselves from this conceptual preparatory work when they envisaged in the Communist Manifesto of 1848, 16 years after Goethe's death, a society which would be "an association in which the free development of each is the condition for the free development of all". Is there an echo of this in the concept of *sustainable development*? Many of the delegates in Rio would have been very familiar with this Marxist definition.

Ernst Haeckel, a contemporary of Marx and Engels, for his part chose the Dar-winian theory of evolution as the key to understanding the world: "'Evolution' shall be from this point on the magic word by means of which we can solve all the puzzles which surround us, or at least start out on the path to their solution."

In such usages, the original meaning of evolution – 'development' – is still present. The image of the opening seed of corn can be sensed in the background. Both terms imply the unfolding of qualities which are inherent. It was the 20th century which first, and tendentiously, made 'development' synonymous with economic growth and industrialisation.

When Haeckel defined *ecology* in 1866 as a science of the relations between the organism and its surrounding external world (*zur umgebenden Aussenwelt*), he was using a circumlocution for *environment*, the second key word at Rio. And another surprise: the coining and subsequent global career of the word *environment*, in German *Umwelt*, are closely linked with the world of Weimar. Both the German and the English terms originated there, though the story is a genuinely European one. The German word *Umwelt* was coined by a Danish poet, whereas the English word *environment* first appears in the translation of a German text by a Scottish writer – put together on the basis of French vocabulary. A very convoluted story, whose fulcrum is Goethe.[14]

> The day is so long, my reflections are undisturbed, and the splendid sights of the environment ('Umwelt') by no means inhibit the poetic sense. Rather, along with the movement and open air, they evoke it all the more quickly.[15]

This is how the modern German term first appears in Goethe's work. In the first chapter of his *Italian Journey* he describes his impressions when crossing the Alps. He wrote the text in 1814-15. A passionate collector of words, Goethe had first come across this one some years earlier in a poem by the Danish lyricist and travel writer Jens Baggesen. It had celebrated its debut in the year 1800 in a long poem of Baggesen's on Napoleon, evidently composed in German.

Born on the Danish island of Zealand in 1764, Baggesen began his career writing ornate verse in the rococo style. He spent many years of his restless life journeying throughout western Europe. On his travels he met with the then avant-garde of German literature. Johann Heinrich Voss, the translator of Homer and Shakespeare, became his friend. He got to know Friedrich Schiller in 1790 during a lengthy stay in Weimar. One year later he rescued Schiller (the literary giant was chronically hard up) from acute money problems by arranging a generous stipendium for him from a Danish prince. In revolutionary Paris he danced "a solo on the ruins of the Bastille". In his poem of 1800, Baggesen writes, "And flood turns into fire, mist into northern lights, / Rain into radiant outpour, so that to the environment (*Umwelt*) / The poet's hellish fate appears as an ethereal castle." The verses deal with the perception of a poetic genius by his environment. What is meant here is the social environment, the public, society.

Goethe comes across this neologism when reading through Baggesen's poems in 1808. Its semantic proximity to the Italian *ambiente* and French *milieu* probably struck him immediately; those words would most certainly have been familiar to him. Now though he uses *Umwelt* himself in a subtly different meaning. "The splendid sights of the environment" – in the 'Italian Journey', this refers to the whole spectrum of impressions received on the journey between Lake Walchen in Upper Bavaria and the Brenner Pass: the elasticity of the air, the cloud masses, the snow-capped peaks of the limestone Alps, the mountain forests of larch-trees and stonepine, and the outward appearance and traditional costumes of the local peasantry. Environment here moves noticeably closer to *ecology*.

The coining of the English word *environment*, too, harks back to a Goethe text. The influence of his surroundings, his *Umwelt* or environment, on the development of his personality became the dominant theme of his autobiography, *Dichtung und Wahrheit* (usually translated as 'Poetry and Truth'), which began to appear in 1811. In the 13th book he describes in great detail the genesis of his story *The Sorrows of Young Werther*, written when he was 24, which deals with the suicide of a young man. He analyses in retrospect the youth culture of those years, when *ennui*, world-weariness and a fascination for the morbid became fashionable. The enthusiastic reception of English literature in Germany around this time had strengthened this intellectual climate. *Hamlet* had become cult reading among young people. The conviction "that the life of man is but a dream" had entered "into many a head". The 'Ossian' cycle of poems, published in 1760 by the Scottish writer James Macpherson, and loosely based on ancient Gaelic ballads, lured young German readers into the realm of Ultima Thule and into the dream world of Caledonian nights where, in Goethe's words, "we roamed about on the infinite grey heath amidst protruding mossy gravestones, looking around us at the grass blown by a chill wind, and above us at the heavily clouded sky. . . ." Then Goethe continues: "In such an element, with such an environment (*Umgebung*) of circumstances . . . harassed by unsatisfied desires . . . we had recurred . . . to the thought that life, when it no longer suited one, might be cast aside at pleasure."

This is how Goethe's sentence appears in the translation by the Scottish writer Thomas Carlyle.[16]

It was the first translation of this passage from Goethe's autobiography into English, and it was included in Carlyle's great essay with the simple title 'Goethe', a 'survey' of Goethe's writings which appeared in the magazine *Foreign Review* in July 1828.

The word which Carlyle used at this point to translate Goethe's *Umgebung* is *environment*. Of course, he used existing linguistic material. The root is the medieval Latin verb *virare* (to turn around). In medieval French, *environs* and *environnement* signify that which surrounds something. Descartes speaks of the things which surround us (*nous environnent*). In English too, its use as a verb became established. John Evelyn, for example wrote of the rural areas which *environ* London. But the noun form *environment* had not existed until now. Carlyle's translation of Goethe is, so to speak, the birth certificate of a word which is today both universal and ubiquitous.

In his lonely moorland farmhouse in the Scottish countryside, Carlyle had Goethe's picture on the wall. In a letter to Goethe he called him his "spiritual father". Goethe in turn thought the British author "admirable", and was absolutely delighted when in October 1828 he received a copy of the *Foreign Review* with his essay. Carlyle seems to have been pleased with his newly coined word. He used it a couple of times in his later writings.

In the early 20th century, the English word began to establish itself in the specialist vocabulary of psychology. A debate sparked off in the USA asked whether human beings were formed more by their heredity or by their environment.

However, it was in its German form that the word gained entry into the vocabulary of ecology and sustainability. It was the Baltic German naturalist Jakob von Uexküll who introduced it there in 1909 in his book *Umwelt und Innenwelt der Tiere*, 'Environment (*Umwelt*) and Inner World of Animals'. In this theory, *Umwelt* has two dimensions. It includes everything which an organism receives or perceives in terms of sensory stimuli from its surroundings (its *Merkwelt*, or 'sense world'), and every effect which the organism has on its surroundings (*Wirkwelt*, or 'effect world'). The environment and the inner world are in reciprocal interaction. The signals from the environment determine the perception of the entire external world and give it meaning. They form the inner world of the organism, which in turn affects the external world through action and changes it. *Umwelt* here becomes a precise technical term of the natural sciences.

How much of this rich tradition was still widely known when the term 'environment' was reintroduced into everyday usage towards the end of the 1960s? In November 1969, *Der Spiegel*, the influential German news magazine, carried the headline 'Apokalypse 1979 – Der Mensch vergiftet seine Umwelt' ('Apocalypse 1979 – Mankind is poisoning its own environment'). In this instance, *Umwelt* was a translation from the English – the article was written by the American scientist Paul Ehrlich. But it made *Umwelt* a fashionable term in Germany, too. The adoption of *environment, ambiente, miljö, Umwelt*, etc. into the global vocabulary was without doubt a significant semantic event. But the word forfeited some of its complexity. When somebody talks about the *environment*, they are inclined to perceive nature only in terms of pollution and destruction. *Mater natura* has been left behind. Rather than seeing every organism as a living being in its own specific *Umwelt*, or *Merkwelt* plus *Wirkwelt*, living nature is devalued, downgraded to a threatened 'setting' or 'surrounding' for *human* life – which is what is assumed to be supremely or uniquely valuable.

CHAPTER NINE

Measuring the forests

The first impression you receive, as you begin the descent towards any of Germany's major airports, is: how green this country is! From whichever point of the compass you approach, as soon as the plane has emerged from the cloud cover, what strikes you first is the lush green of the forests. Many a visitor from abroad will be immediately reminded of Grimms' fairy tales, still today the source of the first images of Germany for many children all over the world. But the closer one gets to the ground, the more one sees – even through the window of an aeroplane – of the ugly side of Germany's landscape. The brutal lacerations made by roads and supply routes are unmistakable. Huge areas have been lost to urban sprawl, cleared, built over and sealed with concrete. And this is an accelerating trend. River beds have been canalised and straightened, deciduous forests have largely been reduced to the status of scattered islands surrounded by monocultural pine forests and agro-steppes. Is there still a vision of sustainability beneath the surface of this landscape? If so, what is that vision?

100 hectares of old-growth forest

My quest led me to the Steigerwald, a small, compact range of hills which rises abruptly from the plain of the Main river. With me was Georg Sperber, a Bavarian forester and an internationally-renowned pioneer of close-to-nature forestry. Since his time as head of the forestry department there, he knows the beech forests of Ebrach better than anyone. Walking with him, I discovered the archetypal ideal of a German forest. It is something you need to experience if you want to know how the idea of sustainability has transformed forest ecology.

Blue sky, icy East wind. On this March morning, my guide and I are tramping over hilly terrain. The natural forest reserve of Waldhaus: 100 hectares of old-growth forest in the middle of extensive, sustainably managed beech forests. We are in the northern part of the Steigerwald, just above the town of Ebrach, a couple of kilo-metres from the motorway between Würzburg and Nuremberg. In this phase of the natural cycle, shortly before the vernal equinox, the sunlight still passes easily between the grey columns of beech trunks and penetrates through to the leaf-strewn forest floor. It has tempted forth the entire palette of early blooming flow-ers. Cowslip, wood violet, hollowroot, blooming carpets of wood anemone, the pale blue kidneywort. Over there in the swampy copse of alders, a stream babbles. In these early spring nights, the female fire salamanders come to the water to spawn, depositing their larvae in the still pools between the moss-covered stones.

Above the tree-tops the kraa-kraa of the ravens can be heard. A wren twitters. 'Silent spring'? Certainly not here . . . The woods are teeming with a stunning variety of different birds. The population density of individual species is unusu-ally high. "Here," says Sperber, "among the species typical for this kind of forest, we have higher numbers of breeding pairs per unit of area, by a factor of four or five, than in younger, cultivated forests with less dead wood." Just before sun-rise, a polyphonic dawn chorus arises all around. From the fluting melodic phrases of the robin to the grunting of the stock (or hollow-tree) dove, from the throbbing drumming of the black woodpecker to the delicate tapping of its lesser spotted cousin, every song is here.

A crazy profusion of life burgeons and romps wild and free beneath an umbrella held aloft by wooden patriarchs some 200 or even 300 years old. It is a tightly interlinked mosaic of habitats. Different forest communities thrive close together on the sometimes sandy, sometimes clay Keuper layers, in different stages of development, from sapling to tree carcass. It is this whole systematic mosaic which produces biodiversity. "Look over there," cries Sperber, "look at the amazing oak next to this fallen beech!" It had been forced to wait for perhaps 100 or 200 years for this gap to open up. And now it was able to expand, and stretch out its crown, and wait patiently for more of the surrounding beeches to die off, and live on for another 600 years.

An abundance of dead wood is the distinctive characteristic of an old-growth forest. From giant split standing trees to the long mound of duff which betrays the remains of a tree returning to the soil, every stage is represented here. Fungal infestation begins on the living tree with the tinder-fungus. Sperber points to a colony of spherical fungi on the moss-covered carcass of a beech. "This is pear-shaped puffball, typical of the final phase of decomposition." Then a hollowed-out tree trunk, apparently burnt black on the inside. Struck by lightning? No; the colour comes from the fruiting bodies, the spore-producing organs, of the carbon cushion fungus. And right next to that area, there are grooves along the trunk: tracks made by the feeding larvae of the hermit beetle. This extremely rare insect colonises hollow tree trunks and is a valuable indicator species for biodiversity.

It is a sure sign of the presence of thousands of other organisms dependent on wood. The true treasure chest of biodiversity is invisible. Micro-organisms produce the air that we breathe and the soil which produces our food. The wood and the forest floor capture the CO_2 which is worryingly changing our climate.

This patch of wood covers no more than 100 hectares. Here, the evolving mosaic of the beech forest has almost reached its climax. Georg Sperber calls the old-growth beech forests "Germany's ultimate natural heritage". Only in reserves like this can the European beech live out its full cycle of growth and decay, its life strategy; only here can the full spectrum of species which depend on it as a habitat develop, a process which takes centuries. The area of protected natural beech forest reserves in Germany now amounts to only 40,000 hectares. This is a minimal, residual natural asset. By way of comparison, the annual area built over and sealed with concrete in Germany is 43,000 hectares. Every year. And it is not slowing down. The global benchmark provides the real shock: between the years 2000 and 2005, the planet lost woodland equivalent to the surface area of the whole of Germany.

"I think our society doesn't realise what it has committed itself to with the Rio declaration on sustainable development," Georg Sperber said to me during our walk through the woods. "It means turning industrial society inside out, right down to its core – a revolution in the most literal sense of the word."

The enchanted forest, and the reality

Roughly one third of Germany's surface area today is wooded. This proportion is high compared with many other countries. More surprisingly, it has remained stable for hundreds of years. It is true that around 1800, at the time of the brothers Grimm, the woods were shrinking rapidly here as well. Perhaps regret for this loss is one reason why the fairy tales take place predominantly in the woods, woods which resemble the old-growth forest in the Steigerwald. Sometimes they are stately and flooded with light, as in Little Red Riding Hood, sometimes dark and impenetrable, as in Iron John. The animals, plants and other creatures which inhabit the woods can be helpful and friendly. But only when they are treated with respect. If not, then the depths of the woods hide deadly dangers. Whatever one shouts into the woods, that is what one hears echoing back. The enchanted forest of the fairy tales always sets the little heroine or hero an existential test. They have to prove themselves in the struggle with the monster. But of course this doesn't take place in a real forest, but in the jungle of the unconscious. The enchanted forest reflects our dreams – and our nightmares. The real woods which the Grimm brothers knew in their own lifetimes were totally different. They consisted mainly of clearings littered with tree-stumps, or else steppe-like juniper heathland. In large parts of Germany the woods had been overexploited and badly damaged by industry and by local farmers. The primeval forest of former times had been plundered and badly thinned-out. By that time, untouched nature, in the sense of not being subjected to continual human exploitation, existed if at all then only on inaccessible steep mountain slopes remote from any settlements and from navigable waterways, or in the enclosed and strictly policed forests which were the hunting grounds for the nobility.

Around 1800, something happened here which in global terms was almost unprecedented. A society roused itself and took the decision to replant its ruined forests, and to search for a means of ensuring that they could be maintained indefinitely. Carlowitz had supplied the missing term; Anna Amalia's foresters in Saxe-Weimar had taken it up. In the decades which followed, the concept was translated into practice everywhere. By methods appropriate to the age of Reason, sustainability was operationalised, mathematicised, economicised and inscribed onto nature. In this way it rose to become the holy grail of a new system of forestry. And by this means it succeeded in stopping and reversing the deforestation. A success story, then?

Yes; but the success story of German forestry is not without its own hidden traps and dangers. All too often, the wood was obscured by the trees.

It is a lesson to heed in the 21st century, when the deforestation of the planet is endangering not only the survival of the billions of people still dependent on firewood for their existence. It threatens biodiversity. It is a massive threat to the climate, the basis on which everyone's livelihood depends. Reversing the deforestation of the Earth is one of the greatest challenges for the future.

To understand what happened to the German woods, one has to keep in mind the images depicted earlier from the old-growth forest.

Working on the term

How was sustainability defined in the 18th century? What kind of vision was associated with it? One of the most creative minds working in this area was that of Johann Heinrich Jung-Stilling, an expert in cameralism, a member of the Pietist movement, an optician and a friend of Goethe's from his Strasbourg days and the *Sturm und Drang* movement. "In spite of an antiquated dress, his form had something delicate about it, with a certain sturdiness." This is how Goethe later remembered his fellow-student. "A bag-wig did not disfigure his significant and pleasing countenance. His voice was mild, without being soft and weak: it became even melodious and powerful as soon as his ardour was roused, which was very easily done."

In his Marburg university lectures on cameralism, delivered between 1787 and 1792, Jung-Stilling, now a professor himself, deals in detail with forestry, and in particular with the measures by which the forest manager can determine how much must be felled each year to ensure that neither are the woods laid waste nor yet that timber is left unused and allowed to go off; that is, that the woods are managed sustainably. Jung-Stilling grew up in a village in the densely-wooded area of Siegerland. As a child he helped his grandfather, an old-time charcoal burner, in his work at the kiln. He knew the shady cathedrals of the beech forests along the ridge of the Rothaargebirge mountains; and he was familiar with the rural economy around the mining villages of the Siegerland, where coppicing had been practised cooperatively for many generations to meet the needs of the villagers and to feed the iron works of the district.

For the German cameralists, it was self-evident that all economic processes were embedded in the cycles of nature. "The earth", Jung-Stilling writes in 1779, "is the true mother of all things which satisfy our needs; she brings them forth and nourishes and sustains them, and eventually she takes their parts, now separated by the process of decomposition, back into her womb, dissolves them and recombines them." In cameralist doctrine, economics was bound up in a social ethics directed towards the general good or universal happiness. Its basis, just as in the Brundtland report 200 years later, was the belief that nobody should be unable to satisfy their essential needs.

Jung-Stilling, too, sees human basic needs as a pyramid. Its base consists of the needs related to the maintenance and continuation of present existence. "These must be satisfied," he remarks succinctly, "because otherwise one cannot live." Above this level come the needs related to the enhancement and perfection of our physical and moral condition. Among these he includes a congenial dwelling and clothing, pleasant furniture, beautiful paintings and music, in short "everything which makes me cheerful and happy and the more ready and able to contribute to the common good". These needs *may* be satisfied; "indeed, as long as the satisfaction of my or others' essential needs is not harmed thereby, they must be satisfied." The apex of his pyramid is congruent with Maslow's levels of self-esteem and self-actualisation. For this grandson of a charcoal burner, who grew up in extreme poverty, needs based on luxury and selfishness are reprehensible, they are – as he expresses it, 200 years before Herbert Marcuse – "false needs". These consist of, on the one hand, those which satisfy only "the taste for pleasure"; and on the other, those which "further only one's own interests, and are detrimental to the common good". Jung-Stilling's theory of needs encompasses provision for coming generations. In his textbook of 1792, he defines the term *nachhaltig* – sustainable – in relation to our dealings with wood as a core resource. "By sustainable harvesting I mean cutting every year neither more nor less than what grows back again, so that the means of satisfying posterity's need for wood is also ensured."

The classic definition of sustainability in forestry was given by a forester from Hesse. In 1795, Georg Ludwig Hartig publishes a textbook with the title *Guide to the Evaluation of Forests, or to the Determination of the Yield thereof in Timber*. At this time, Hartig is 30 years old and in the service of the Prince of Orange-Nassau at the family seat of the dynasty in Dillenburg. When the Prince comes for his annual hunting trip on the local hunting grounds, Hartig dons the gala uniform of the corps of foresters for the reception: a green jacket, white breeches, black boots, a plumed hat, and a cutlass hanging from his belt. Hartig comes from a region shaped by extensive deciduous forests and by the coppicing economy of the iron smelting industry. He was born the son of a forester in Gladenbach near Marburg and received his early training in forestry from an uncle in the Harz mountains. In the second edition of his book, which appeared in 1804, Hartig offers his frequently cited definition: "No long-term forestry can be imagined or expected, if the yield of timber is not calculated with regard to sustainability (*Nachhaltigkeit*)."[1] Every wise forest manager must therefore immediately evaluate the public forests for their maximum yield, but must seek to utilise them in such a way that posterity derives at least as much benefit from them as does the present generation. For foresters of the period around 1800 there was one iron rule: the upper limit to the use of natural resources is determined by the natural rate of regrowth. The foresters in the various principalities knew very well that the trees will not grow any faster just because the Treasury needs a few thousand Gulders more per year.

How was the word *Nachhaltigkeit* defined in contemporary German dictionaries? Let us take a look at the *Wörterbuch der deutschen Sprache* (Dictionary of the German Language) published in 1809 in Brunswick. The editor was Johann Heinrich Campe. Campe was not an unknown. He had taught the Humboldt brothers in Berlin and later translated Defoe's *Robinson Crusoe*, and had been an avid supporter of the French Revolution. In his dictionary he defines *Nachhalt* as, literally translated, ". . . what one holds on to when everything else does not hold any more." In other words, *Nachhalt* is what sustains you when everything else is not sustained any more – and collapses. Remember how the modern term appeared in the 1972 *Report to the Club of Rome*? Its authors were looking for a world model that was "sustainable, i.e. without sudden and uncontrollable collapse". In both cases 'sustainable' is introduced as the antonym, the opposite of 'collapse'. This definition surely opens a door to the profound, the inner meaning of the concept.

Barely 100 years after the appearance of Carlowitz's *Sylvicultura oeconomica*, the idea and the terminology of sustainability have put down roots in the German-speaking world. The guiding principle is the rational, mathematical organisation of the forests. In the spirit of Descartes and Gauss, 'geometric operations' are the key to the eradication of over-exploitation. The sovereign lords of the German princely states, and their master foresters, consider themselves to be 'maîtres et possesseurs', the 'masters of Nature'. The programme of reafforestation within the German territories proceeds on this intellectual basis. 'Establishing new stocks' is the contemporary expression. "The cultivation of young forests" is intended to guarantee "the perpetual duration of the woods." (Hartig) This generates a self-contained hierarchical bureaucracy, and a branch of the research and education industry, which attract worldwide attention by the early 19th century. The forestry academy in the Saxon town of Tharandt becomes the Mecca of the international science of forestry.

The academy for sustainability

'Site of memory': Tharandt. The journey from the main railway station in Dresden takes just twenty minutes on the S-Bahn, the rapid urban train system. You leave behind the metropolitan sea of houses, the industrial zones, the housing estates and the allotments. The rail tracks follow the course of the Wild Weisseritz river. A small floodplain wood, a meadow, then the chimneys and hangars of a steelworks. The valley narrows. The slopes to left and right are now wooded. Next stop: Tharandt. The sign reads 'Forest Town Tharandt'. This small town at the edge of an extensive forest is the seat of one of the oldest functioning forestry academies in the world, and for a long time the most prestigious.

Its present-day name is The Faculty of Forest, Geo and Hydro Sciences of the Technical University Dresden. The institute rose to global fame under its old name of Royal Saxon Academy of Forestry at Tharandt. This is where the sum of human knowledge about the forests was stored and enriched, and where the idea of sustainability was scientifically explored, elaborated and exported around the world.

The buildings of the Faculty stand in a neat row in the valley of the Weisseritz river: modern eco-architecture in wood and glass, prefabricated concrete constructions from the GDR period, stone cubes adorned with plaster embellishments from the Biedermeier period of the first half of the 19th century. A steep path leads up past the ruins of a fortress to the entrance to the botanical forest gardens. Spread out over this huge area the visitor can find close to 1,500 types of woody plant from all over the world in natural setttings. From the Dawn Redwood and the Swamp Cypress, which laid down in earlier geological periods the fossil fuel deposits, to exotic species like the Sugi or Japanese Cedar and the Indian Bean Tree. Some individual trees are by now 200 years old. Laying out and planting the arboretum was one of the first things to be undertaken once the lectures had started. The founders of the forestry academy wanted to combine the theory of silviculture with visual examples and to let their students see the thousand active forces of nature with their own eyes.

In 1811 the Thuringian forester Heinrich Cotta, together with a small team of colleagues, took over a former bath-house on the banks of the Weisseritz and founded what was at first a private Institute for the Education of Foresters and Hunters. The king of Saxony had brought him to the country to survey its forests and to manage them. The methodical basis of their management was to be their long-term re-organisation, in space and time, on geometrical and mathematical principles.

"I am a child of the forest," Heinrich Cotta said of himself. His parents' house stood in Zillbach, a village in the foothills of the Rhön mountains. His father belonged to that generation of foresters who had implemented Anna Amalia's reforms. Cotta, born in 1763, began his career as a hunter's apprentice. That was in that same September of 1780 in which Goethe scribbled his Wanderer's Night Song on the board walls of the hunter's hut on the Kickelhahn. Later, he studied mathematics and cameralism for two semesters in Jena. After that he gained practical experience as a land surveyor before joining the Weimar forestry service. His credo, as expressed in 1790, was "to follow Nature, who acknowledges no laws, but allows us to attend her and observe her own". The echoes of Goethe's conception of nature are unmistakable. And conversely Goethe was a great admirer of Cotta's study of plant growth, which appeared in Weimar in 1806, and remained in contact with him until his death. Cotta focused his attention on the growth of trees. How does it proceed? How can the yearly increase be measured? How calculated in advance? And above all, how can growth be accelerated? At that same period, the young philosopher Schelling was lecturing at the university of Jena on Spinoza and the original productive force of nature, the *natura naturans*. For Cotta, this was too far removed from practice. His overriding aim was to discover the maximum yield which can be sustainably obtained from a particular woodland by means of a cultivation plan drawn up specifically for it.

'Managing' the forest

Between 1811 and 1831, while running the academy in Tharandt, Cotta carried out what was called a *Forsteinrichtung* – that is, compiling a comprehensive inventory and drawing up a long-term management plan – for all the forests in the kingdom of Saxony. How was this carried out?

It starts with a walk in the woods. A human chain moves across the terrain, made up of uniformed forestry employees 20 metres apart. Their equipment is similar to a surveyor's: a measuring chain, calipers for gauging the diameter of trees, a hypsometer for measuring tree height, a spirit level, an angle compass with tripod, a drawing board. The boundaries of the property are marked out once again, and distinctly. The total area is carefully measured. Each person involved notes down the number and species of all the trees in their section. The height of the trees is established by a simple trigonometric procedure. At exactly 1.3 metres, around chest height, the calipers are applied in order to measure the diameter of the trunk. The position and dimensions of clearings, where the tree cover has been removed, are also recorded. Sofl quality and climatic conditions are assessed.

Goemetrical and mathematical ingenuity is now applied to calculate the volume of wood in the property, its stock. This quantity becomes the starting point for forecasting and planning. By means of yield tables showing average growth by species for soils of all possible grades, the total amount of additional wood which can be anticipated from the managed area (or 'stand') is now calculated as accurately as possible.

The decisive step is determining the length of the rotation period. This is about growth. Rotation is basically the period of time that elapses between the formation of a stand and the time of its harvesting. This is where forestry management meshes with the rhythms of nature and its growth cycles. When will the trees in a given stand achieve their maximum rate of growth? When will this begin to slow down? The rotation period should end after that point has been reached. At what point will the wood of the trunk begin to decay? The felling should begin well before then. The aim is to achieve the maximum yield when harvesting the timber, to identify the optimal moment for felling the tree.

The determination of the rotation period of a forest links the rhythms of nature with the periodicity of human economics and the laws of supply and demand. What kind of wood, of what quality and quantity, is needed for the local economy? Determining the rotation period has always been a balancing act. On the one side are the ecological factors, namely growth rate and regeneration. On the other side the economic interests bear down: maximum yield, maximum profit.

The word 'rotation' originates in the vocabulary of mechanics. There it describes the circular movement of a mechanical apparatus, which can be adjusted, slowed down or speeded up at will. Determining the length of the rotation period for a forest often followed similar lines of mechanistic or managerial thinking. Financial expectations prevail.

Once the woodland has been measured, the stock of wood and the growth rates have been assessed and the rotation period determined, then the sustainable management of the forest is theoretically within reach. The total available woodland is parcelled up into the same number of felling areas, or 'coupes', as the rotation period has years. Every year, felling moves forward from one coupe to the next. When the last coupe has been cleared of trees, the same number of trees should in theory now be ripe for felling in the very first area. Sustainability has been achieved.

The great unknown in all of this is regeneration. Ensuring that the wood regenerates itself on every cleared area is the main duty of every forester. The German foresters were well aware that the wealth of a country lay not just in its stock of timber, but in the long-term productive capacity of those forests. A sustainable yield of wood is only possible if the vitality of the forest is maintained and strengthened. The essential difference between forestry and industrial production lies in the fact that in the first the source of the raw material is the living natural world. The product (timber) and the means of production (trees) are identical. The harvesting of the product destroys the producing organism.

There are two possible routes to regeneration. In artificial regeneration, new stock is sown or planted on a cleared area. The alternative is natural regeneration. The stock is renewed by the generation of seeds which have fallen or been carried from the surrounding trees, or by vegetative reproduction through root sprouts and stump sprouts. The forester can support this process by encouraging strong crown growth, which aids seed production, by working the soil, which helps seedlings to flourish, and by clever management of the distribution and penetration of light in the forest. The young trees grow bigger under the protective umbrella of the old. At the appropriate time the forester ensures they receive enough light. Natural regeneration is both more economical and more ecological. The forest maintains itself. The new generation of trees is optimally adapted to its location. The German foresters of the 19th century were famous for their skills in natural regeneration.

The 'normal' forest

Trying to manage nature is not unproblematic. Every forest management plan represents an intervention, of whatever severity, in an ecosystem. It imposes a new and artificial spatial and temporal order on a forest. Forest ecosystems are highly complex living systems with a variety of elements: trees, plants, animals, micro-organisms, soil, water, air, etc. The total stock of timber in a large area of woodland can only be estimated very imprecisely, and this is even more true with regard to its growth rate. Natural vegetation allows only approximate planning. There are too many irregularities and deviations, too much uncontrolled growth involved. Nature knows no mathematical regularity and no geometrical forms. No leaf matches any other. No tree grows to be identical to its neighbour. Every piece of woodland is exposed in a different manner to sun, wind and rain, and dependent on soils formed by very specific geological factors. Meteorological events are unpredictable. The small differences and imponderables add up. Forests are difficult to comprehend. In order to make them at all calculable or predictable, they have to be rendered more abstract. Cotta and his followers developed abstract models of an ideal of sustainability on the drawing board in order to help them realise their plans. They invented the 'normal forest'.

A normal forest is made up of 'normal trees'. The trunk of a normal tree consists of a truncated cone at the lower end and a truncated paraboloid at the upper end. It is therefore calculable. If one knows the diameter and the total length of these geometric forms, the volume of timber present in the normal tree can be calculated. Using measurement series taken over many years, the average yearly growth rate for every tree species in every habitat has been recorded and preserved in 'wood growth rate tables' and growth curves. With these yield tables it is possible to establish likely annual growth rates and the optimal age for felling. From this the rotation period and thus the number of felling areas can be derived. A normal forest consists of normal trees of a single species. The soil quality is exactly the same throughout. Skidding trails divide the terrain into equal compartments. Each compartment contains only trees of the same age, all one year older than those on the preceding compartment. On an area of – for example – 100 hectares of normal forest, the first compartment will contain trees which are 99 years old and the last compartment will contain one-year-old seedlings. Felling moves forward from one compartment to the next each year, and seedlings are grown again on the clearfelling. In this way, one compartment is used every year. When after 100 years the final compartment is cleared, a new stock of 99-year-old trees will be ready on the first compartment. The rotation begins again from the beginning. The normal forest is a model of the 'perpetual forest' and therefore absolutely durable and 'sustainable' But it is not 'normal' in the sense that it naturally grows in this way. It is by no means the forest we are most familiar with. It is in fact never found anywhere in nature.

The word 'normal', in the language of the 18th century, does not mean typical or common. The Latin root word 'norma' signifies a right angle, and then by extension anything that is 'regulated'.The normal forest is a totally regulated and standardised forest: a model, a fantasy.

Cotta and his followers, in Tharandt and at the other forestry academies, were under pressure to build up the new forests quickly. The reafforestations and afforestations were carried out using fast-growing tree species such as spruce or pine, and as pure stands or monocultures, in compartments of the same age and class. As this strategy proved successful initially, it was decided that many of the still extant old-growth deciduous woods should be changed to coniferous plantations. The mosaic of the woods turned into the checkerboard of the managed forest. The new forests were basically wood plantations, in which trees stood in military rows grouped according to age. Everything was strictly regulated by human hand, by the purposive and rational human mind. A piece of the living biosphere was redefined as a resource repository and reconstructed accordingly. It is true that this spelled the end of unregulated felling and of over-exploitation, and put a stop to deforestation. But it came at a high cost: the de-naturing of the woods. And it happened in the name of sustainability.

Early warnings

What Elias Canetti in the 20th century identified as the "deep and secret joy" of the Germans in their orderly forests with their straight parallel rows of trees in their clean and clearly defined spaces, is not principally the product of an archaic 'instinct for the forest' or even of a deeply ingrained desire for military discipline. It expresses rather the 19th-century ideal model of sustainable forestry and a deep need for security of supply. The classic authors of German forestry were fully aware that this model was an emergency measure, a stopgap which was not practicable over the longer term. At the unspoken level at least, there was a consensus that sooner or later there would have to be a return to the natural forest.

Cotta himself had urged caution: "The science of forestry cannot work magic and cannot go against the course of nature." His colleague Emil Adolf Rossmässler, a professor of zoology and an admirer of Humboldt, warned that "nature is not a storeroom". Gottlob König, Cotta's colleague and brother-in-law, an interlocutor of Goethe's and a classic forestry author in his own right, expressed the secret credo of the German guild of foresters in 1840 at an assembly in Brno in Moravia (today in the Czech Republic).

> Let us remain faithful to our high calling and not lose sight, in our pursuit of the most productive cultivation of timber, of the natural destiny of the forests. We must cease treating these delicate living beings as if they were mechanically self-regenerating stockpiles, which require only to be split with the axe according to instructions handed down by shortsighted bookkeepers; for the growth and health of the forests hangs from very fine, deep-hidden threads. We still understand only very little of the secrets of nature, and our wisdom, by means of which we propose to pass the management of the woods down from generation to generation in formulas and figures, will be confounded by the crop failures and bad harvests which are nature's revenge for every maltreatment . . . .

Gottlob König was making a plea for forest management in accordance with nature for the sake of the habitability of the earth. This is how he expressed the legacy of his generation: "[W]here the forests and the trees disappear, they are followed by desert and drought . . . without a firm bed, flash floods carve trenches through the land, winds drive the soil up and around in the air and destroy every new germ of life. The fall of the first tree was the beginning of civilisation, as is well-known, but the fall of the last will just as certainly be its end. We move between these two poles of human life on earth. The time when we will reach the last of them lies in our hands."

A prophetic warning for the 21st century, when turbo-forests, short rotation plantations and monocultural energy crop cultivation are expected to produce biomass to feed the world's hunger for energy – 'sustainably'.

The invisible hand of the market

But the zeitgeist – a word coined by Herder – was blowing from a different direction. "Peter Munk had now reached the highest point of the pine grove, and took his stand before a tree of prodigious girth, which a Dutch shipwright would have given many hundred florins for as it stood." A fairy tale, written in 1827 by Wilhelm Hauff. The cold heart of capitalism was not to be reconciled with traditional sustainability thinking. In the middle of the 19th century the clash came to a head. The stage was once again the forestry academy at Tharandt.

It was there in 1865 that the forestry mathematician Max Robert Pressler attempted to establish 'soil rent theory' as the new guiding concept. The majority of foresters vehemently rejected it. A bitter struggle for supremacy broke out which still continues today.

Soil rent theory represented a radical break with traditions which went back in principle to the Venetian forest regulations of the 15th century. This new approach was based on establishing the net profit, that is, the highest possible return on the capital invested in the forest. How can silviculture be organised so that investment produces a financial return at least as high as that of any other possible usage, for example conversion into arable land, or as a location for manufacturing? In place of the maximum steady yield in timber, suddenly it was the maximum monetary yield of a wooded area which was at the centre of decision-making in forestry. The guiding principle became the 'sustainable' – i.e. long-term – maximum return on capital. Meeting the society's long-term needs for timber had traditionally been the highest aim and paramount duty of forestry. The new guiding principle put the focus on the financial returns to the public and private owners of the woodlands. The benchmark was no longer the productivity of nature but the free market and its laws of supply and demand. The rhythms of nature were overlaid by the dynamism of capitalism, utility value by exchange value. Sustainability was uncoupled from nature and from society's need for natural products. The principle of responsibility for coming generations gave way to the forest owner's long-term cost-benefit analysis. A system change was in the offing, under the umbrella of a term, a label, which remained the same. The combination of the 'normal forest' model with 'soil rent theory' brought about deep changes in forestry. An abstract theoretical model was transposed onto actually existing forests. Reduction of the rotation periods and the conversion of the woods into coniferous plantations were the tools which were now applied to bring this about.

The vocabulary of sustainability was by no means abandoned at this point. It was only loosened or redefined. Now, a simple constant, steady monetary profit, or the retention of a wooded area for forestry use, was considered a sufficient criterion of sustainability. "With the different unconnected definitions available," the Prussian forester Bernhard Borggreve mocked in the 1880s, "one can do little, or – if one prefers – everything." Thus, even the most extreme forms of pillage can be euphemistically defined and defended as sustainable. Other foresters asked openly whether sustainability was to be defined in future by 'the credit institutes'.

The 'invisible hand of the market' seemed for a while to be steering things in a positive direction. When railway construction, the paper industry and coal mining created a huge demand for cheap softwoods in the second half of the 19th century, the German forest owners were able to supply the market. Profits rose to undreamt-of heights. Then, however, the invisible hand of nature came into play. It was not long before nature's revenge for its maltreatment, of which Gottlob König had warned, made itself felt. The new forests proved to be extremely unstable. Monocultures are never sustainable. In the 1850s already, the caterpillar of the nun moth laid waste the coniferous forests of East Prussia. This was the cue for the emergence of a new breed of experts. The entomologists now took the stage. The 'war against nature', which Rachel Carson was to write about 100 years later, was escalating.

The invention of the pest

Up to that time, it had not occurred to anybody to question whether insects belonged in the woods. The calamities which they brought on from time to time were regarded as punishments from God. A contemporary of Carlowitz, the Thuringian Pietist Friedrich Christian Lesser, was able to note coolly in 1738 how "by means of attentive observation of the insects, to whom we usually pay little regard, a person can attain a living knowledge and veneration of the omnipotence . . . of God on high". In Heinrich Cotta's first curriculum for the forestry academy at Tharandt, the topic appears under the neutral title of 'Natural History of the Insects of the Woods'. And in 1834 Rossmässler was still writing dispassionately about "the insects which cause most damage to the tree species cultivated here". But only three years later, in 1837, Julius Ratzeburg, Professor at the Prussian Forestry Academy at Eberswalde, published a book about *Forest Insects* which bemoaned the "harmful insects" which "change in any way the normal condition of woody plants". In 1841 he followed this with the standard work, *The Destroyers of the Woods, and their Enemies*. The tone had become much more aggressive. Ratzeburg recommended that plant-eating insects which damage the woods should be combatted through the encouragement of their natural enemies – songbirds, and the parasites which prey on the unwanted insects. This recipe was very much still in line with classical thinking.

But the calamities in the monocultural 'normal forests' grew ever more frequent, and thwarted the expectation of 'sustainable' profits. At the same time, the ideas of Darwin and his followers about the 'struggle for existence' were conquering the natural sciences. Often interpreted one-sidedly, they nevertheless gained a foothold in forest biology as well. In the terminology used here, the 'harmful insects' have now become 'pests'. At the end of the 19th century, the ant expert Karl Escherich established himself as the master of the dark arts of pest control. In 1907, in the venerable lecture hall of the Forestry Academy at Tharandt, he held his inaugural lecture on 'The elemental force of reproduction'. In a radicalised Darwinian terminology he declared war on "the abnormal proliferation of pests". His thinking focused unwaveringly on the "mass destruction of verminous insects" by any conceivable biological, toxicological or technical means of mass extermination. The biological arsenal was soon augmented by the addition of chemical weapons.

A few years before the Tharandt academy was founded, in 1797-98, Francisco de Goya painted a famous cycle of pictures in Madrid. One of them shows a man bent over a desk. His head is resting on his arms. He is sleeping. Towards the back the picture gets darker. From there a swarm of night creatures flies out, as if they want to attack the sleeping man. They are hybrid dream animals, owl-like birds, oversized bats. One animal, part wildcat, part sphinx, fixes its gaze on the sleeping man. The front of the desk bears an inscription: *el sueño de la razón produce monstruos*. The sentence is ambiguous. For *sueño* can be translated either with 'sleep' or with 'dream'. The sentence can thus mean either: the sleep (= absence) of reason produces monsters; or, the dream (= fantasy) of reason produces monsters. It is precisely this ambiguity which makes the picture a parable of the Cartesian project of controlling nature by means of instrumental reason.

Was there an alternative course to that of 'rational' forest management? Around 1900, the German forestry community was arguing passionately about a way *back* to nature. Substantial ideas were emerging about a silviculture in accordance with nature. The idea of the 'permanent forest', which conceived of the forest as a living organism, was gaining ground against the model of the 'normal' forest. Interest was growing again in the processes which take place in virgin forests. Seeing how nature does it, and giving priority to the full potential natural vegetation, and thus to biodiversity, were the new watchwords. The rational imperative now seemed to be to bring together the principle of sustainability and the idea of ecology. However, the context had changed dramatically. The economic pressure on the forests had now declined. The fossil fuel era was already in full swing.

CHAPTER TEN

Fossil, nuclear, solar

Oleum – the Latin word for olive oil – was in medieval English the basis for the formation of the word 'æle'. Until around 1300 this word exclusively meant olive oil. Then it extended its meaning to include oils extracted from hempseed, flax, poppies, sunflowers, beechnuts or walnuts. But it was always a vegetable raw material, one which replenished itself. The use of the word 'oil' for 'mineral oil' or 'petroleum' (a combination of *oleum* and the Latin word for rock, *petra*) was first recorded around 1520, but was not common until the late 19th century. The word 'coal' originally meant charcoal, produced in kilns by heating beech, birch and other types of wood. Fossilised carbon was called 'sea-coal'. This term dates back to the mid-13th century. The principal source of this type of coal at that time was underwater deposits washed out into the North Sea and collected in pieces on the North Sea beaches. The term 'sea-coal' was retained in later times when it was extracted in collieries from underground fields around the towns of Newcastle and Durham and transported by sea to the London region. It became the dominant meaning of 'coal' as sea-coal increasingly came to replace firewood and charcoal during the 17th and 18th centuries. On the European mainland, the semantics of 'coal' changed much more slowly. The Grimm Brothers' German dictionary mentions in volume 11, published in 1873, that "only since recent times *Kohle* (coal) signifies black coal rather than charcoal".

This change in language reflects the fact that our civilisation was based from its beginnings until the industrial revolution almost entirely on self-regenerating raw materials and the renewable energy sources of biomass, wind and water. The only additional source of energy was the muscle power of people and animals. All of these resources are constantly being renewed, through the movement of the air, the circulation of water, or vegetable photosynthesis. Only the constant influx of sunlight keeps the *oeconomia naturae* going. Non-renewable metals and minerals in the earth's crust were being used, it is true – but only very sparingly. Apart from that, humanity lived on sunlight alone. There was as good as no waste. Everything disappeared again in the great circulatory system of nature. The great change occurred with the large-scale switch to fossil fuels.

Fossil fuel deposits are also of course the products of solar energy. But their generation required geological passages of time, meaning millions of years. Countless generations of forests, consisting of tree-like clubmosses, horsetail and other plants, died off, were buried beneath heavy stone and carbonised, and were deposited at many points of the Earth's crust as coal, mineral oil and natural gas. People had always known about these deposits and their qualities.

In the China of the Song period (between 1000 and 1200), coke was already being used instead of charcoal for the smelting of ores wherever wood became scarce. The conquistadors certainly knew about the mineral oil deposits in South America. In 1637, Alvaro Alonso Barba, a cathedral priest in the Peruvian silver city of Potosí and an authority on mining, wrote: "Naphtha, or mineral oil . . . is a sulphurous liquid matter . . . which has wondrous powers. It draws on the fire . . . with such power, that it will take fire even when at a fairly great distance from the flame." But the exploitation of the underground deposits took place only on a very restricted scale. Extraction was difficult. Moreover, coal was considered inferior to charcoal. As long as enough trees grew back and the price of wood remained affordable, it wasn't worth the trouble. The changeover occurred late, hesitantly, and always under the pressure of acute shortages of wood.

A surprising discovery is that the use of fossil fuels was initially only a strategy for the protection and preservation of the forests. The intention was to bridge a temporary 'gap' until the transition to a sustainable use of wood was completed. It was in this expectation that the surveying and measurement of the 'subterranean forests' began. That this would lead to a 'fossil fuel era' was suspected by hardly anyone, and desired by no one. Today, nuclear energy is supposed to bridge the 'gap' until the solar age has finally and conclusively arrived. But history shows that 'bridging technologies' can develop their own prodigious momentum.

Wood shortages and 'sea-coal'

It began in the British Isles, which are relatively poor in forests. The Roman occupying forces had already dug for coal there. But this mining activity ceased in the Anglo-Saxon period. The long, slow ascent of the new resource began only as the Middle Ages drew to a close, and at first only on a small local scale. Digging for coal was a craft, practised by peasant farmers, often working for monastic landowners. Coal was first used in village smithies. The driving force for the replacement of firewood by coal was always a shortage of wood, and the mechanism was the market price. Where there was a lack of wood, the price went up. Where coal was available, and cheaper than wood, the substitution occurred. Soon it was also being used for burning lime, boiling salt, brewing beer and bleaching linen. The market grew in leaps and bounds. In London there was great demand for 'sea-coal' (you can still find Newcastle Close and Old Seacoal Lane in the ancient centre of the city today, marking the end point of its long journey by sea and river). The trade, and subsequently also the extraction itself, was concentrated in the hands of a powerful cartel, the 'Lords of the coal'. The region around the River Tyne was called 'new Peru', in an allusion to the silver-rich Spanish colony.

Historians of Britain place the breakthrough for the new resource and its associated technologies in the Elizabethan period. That was the second half of the 16th century, the era of the beginning of colonialism. It was the age of Shakespeare, whose Prince Hamlet famously worried that "the time is out of joint". Between 1530 and 1650 the population of the islands doubled. The spread of arable land and the increasing rate of tree-felling accelerated the deforestation. As early as 1700, less than 10% of the land area was still under trees or hedge shrubs. The final breakthrough for coal came when the populace of London – at first the poor, then also the middle classes and the nobility – began to heat their homes with coal. It is estimated that around 1650 more than half of the energy needs of the kingdom were met from fossil fuels. For the first time in the history of the planet a non-renewable raw material had become the principal energy source for a whole country.

The fossil fuel revolution did not take place without resistance. The problem of emissions grew virulent very quickly. It was John Evelyn, the 'virtuoso' whom we already know from his role as a pioneer of the philosophy of sustainability in Europe, who placed himself at the head of the protestors in England. In 1661, four years before he called for a national campaign of tree-planting in *Sylva*, his pamphlet *Fumifugium* appeared. In it Evelyn attacks the "pollution of the Aer" caused by "the immoderate use of, and indulgence to Sea Coal in the City of London that . . . so noble, and otherwise, incomparable City."[1] He asks why "our goodly Metropolis should wrap her stately head in Clowds of Smoake and Sulphur, so full of stink and Darkness". And he curses "that Hellish and dismall Cloud of Sea Coal which is . . . perpetually imminent over her head." The noxious smoke from the chimneys coats everything with a sooty crust. "There is nothing free from its universal contamination." Evelyn cites the rain, the dew, the leaves of the trees, the orchards and their fruit. It contaminates everything which is exposed to it. "This horrid Smoake, obscures our Churches, and makes our Palaces look old." The emissions from only one or two chimneys not far from Scotland Yard invaded even the rooms and galleries in the Palace of Whitehall; the Duchess of Orleans, the King's sister, had complained bitterly to him (Evelyn) about this. "The City of London resembles the Suburbs of Hell rather than an Assembly of Rational Creatures."

The "Arsenical vapour" rising from the "subterrany Fuell", which endangers the plants and the buildings, jeopardised the health of the people as well. "In the London churches and Assemblies of people . . . the Barking and Spitting is uncessant." The inhabitants "are never free from Coughs and importunate Rheumatisms". The symptoms: "loss of Appetite, Cathars and Distillations". The consequences: "Almost one half of them who perish in London, dye of Phtisical and Pulmonic distempers." The fault for all of this lies in the excessive and reckless use of "New Castle Coale" in the industrial workshops of the city.

At that point Evelyn turns to discussing the connection between affluence, greed, the environmental crisis and the "Publick Good". He expresses his amazement that "where there is so great an affluence of all things which may render the People of this vast City, the most happy upon Earth; the sordid Avarice of some few Particular Persons, should be suffered to prejudice the health and felicity of so many." He goes on to explain that "it is not happiness to possess Gold, but to enjoy the Effects of it, and to know how to live cheerfully and in health." And he attacks the indifference of his contemporaries towards their own wellbeing: "That men whose every Being is Aer, should not breath it freely when they may; but. . . condemn themselves to this misery. . . is strange stupidity." He quotes an epigram of the Latin poet Marcus Valerius Martial: *non est vivere, sed valere vita*. Freely translated, Martial's phrase means: life is not simply about being alive, but being in good health and in full possession of one's faculties. Reproduced on countless websites on the internet, this aphorism remains a popular key to the art of living today.

Evelyn's proposal was that the factories and manufacturers polluting the air, such as brewers, salt-boilers and lime-burners, should be "removed five or six miles from London below the River of Thames", thus creating a green belt for the protection of the urban population. He also makes a plea – as he did a little later in *Sylva* – for the planting of trees and of fragrant herbs in the gardens of the city. Last but not least, he urges the importation of firewood from the Nordic countries instead of sea-coal from Newcastle. The book met with a hostile reception. Medical experts from the "Colledge of Physicians" questioned Evelyn's findings. In their judgement the smoke from the Newcastle coal might even help "the Preservation against Infections". Economists argued that "to talk of serving this vast City with Wood, were madnesse".

With the onset of the fossil fuel era, a new vocabulary emerged. The 'gifts' of (according to choice) God or Nature now became 'resources'. The linguistic change reflects a change of perspective. The word comes from the Latin *resurgere* – to rise again, to spring up again. It first appears in English around 1600. The word 'resources' calls to mind the fetching of materials which had been stored away, the mining of underground deposits. Only a hundred years later it was being used to signify all of a country's original sources of wealth. Karl Marx, who was well versed in the language of the classical British economists, wrote that "capitalist production . . . develops technology, and the combining together of various processes into a social whole, only by sapping the original sources (*Springquellen*) of all wealth – the soil and the labourer." The gifts of nature are essentially a means of living; resources are often only a means of making profit.

The subterranean forests[2]

The 'subterranean forests' were discovered on the Continent, too. The metaphor was used in 1693 already by the Brandenburg cameralist Johann Philip Bünting, a contemporary of Carlowitz. Through the "wondrous providence of all-merciful God", he wrote, mankind has been "granted the gift of a *sylva subterranea* or underground forest of coal". With this gift the society could bridge the gap in supply which had arisen through the decline in "wood that grows wild". In 1765 the phrase appeared in a pamphlet of the Economic Society of Berne. The author was the Polish Count Joseph Mniszech. He wrote concerning the origin of fossil fuels that "many people ascribe these to the Great Flood; thence the underground forests, which lie buried in deep peat bogs".

Every crisis, every spike in the ongoing shortage of wood gave further impetus to the search for ways of reducing its use. Carlowitz had already pointed to the peat deposits of the high moorlands in 1713. His successors in the administration of the Electorate of Saxony initiated the systematic mining of 'earth-coal' there.

One of the pioneers was Johann Gottfried Borlach, who had started his career at the Mines Inspectorate in Freiberg under Carlowitz. In 1724 he was asked to put salt production in the Electorate, based on saltworks along the Saale and Unstrut rivers, on a proper footing. Because of the shortage of wood and charcoal for salt-boiling there, Borlach began a systematic investigation of the prospects for using fossil fuels. In 1738 he went on a study trip to England. The new knowledge with which he returned was that "there it is coal which has the potential to replace wood throughout the land". It was as a colleague of Borlach's in the little Saale town of Kösen that Friedrich Anton von Heynitz began his career in 1742. Subsequently he took over as the director of mining in the lower Harz region, and moved from there to Saxony, where in 1766 he was one of the founders of the Mining Academy in Freiberg. In 1776, the year when Adam Smith's seminal book appeared, Heynitz took a journey through the British industrial centres. The following year he was brought to Prussia by Frederick the Great. In Berlin he became Director of Mining and Minister of State. In these offices he devoted himself principally to the development of the coal-mining industry in the new Prussian province of Silesia and in the county of Mark (which later became the Ruhr area).

Under his direction the first steam engines came into operation in Prussia. These technical marvels of the modern age could not operate to their full potential if fuelled by charcoal.

A spotlight is thrown on the dawn of the fossil fuel economy by a report written in 1792 by Alexander von Humboldt, then a young assessor in von Heynitz's department. Humboldt was 22 years old. After briefly studying cameralism at the University of Frankfurt an der Oder, he had travelled to England with Georg Forster (who had cicumnavigated the globe as a member of one of James Cook's expeditions), not least in order to inspect the coal mines there. After that he underwent nine months of highly intensive practical education in geology and mining at the Mining Academy in Freiberg. Now the freshly-qualified Prussian mining expert faced the first substantial task of his career. He was to inspect the mining and metalworking industries in the principalities of Ansbach and Bayreuth, which had recently been absorbed into Prussia, and to report in detail to the Minister for Mining in Berlin, von Heynitz.

In the summer of 1792 he ranged on horseback through the valleys and hills, the lonely woods and moors of the Franconian forest and the Fichtelgebirge mountains, and inspected at first hand the small mineshafts, tunnels and metalworks where modest quantities of iron, copper, cobalt and – not far from Bayreuth – even gold were extracted and smelted.

In one of his first reports to Berlin, Humboldt speaks – as Carlowitz did 80 years before him – of the shortage of wood, which could jeopardise the extraction and smelting of ores in the longer term. Although the locals still believed in the inexhaustibility of the woods, the signals of an imminent shortage of timber should be taken seriously. Rising costs for wood could jeopardise the hoped-for profitability of the mines. The only fuel which is used to date is wood. Improved silvicultural methods, however, had only very recently been introduced, or were still at a very early stage. "A division into regular felling areas, and the accurate forestry surveying and measuring which must precede it, is still entirely absent, and without these a steady and secure forest management, a balance between regrowth and consumption (*Consumo*), is hardly conceivable." What Humboldt defines with precision here is sustainable forestry, just as it was envisaged in these years around 1800 by Hartig and Cotta. Immediately following this, however, the young Humboldt draws attention to an alternative fuel. He has observed, in the vicinity of Kronach, that small quantities of coal were being used in a smithy. In addition, on his travels he has seen several areas of peat moorland which could also be cut for fuel. He recommends that these fuels should be used for the mines and metalworks in the principality of Bayreuth in order to bridge the gap of 20-30 years before the improved forestry methods were fully established. Humboldt of course could hardly have foreseen that 200 years later the bridging of a 'gap' in energy supply would once again be invoked, this time to justify the continued use of nuclear power.

Eight years after Humboldt's explorations in the Fichtelgebirge mountains, in the early summer of the year 1800, the Saxon state mining expert and saltworks inspector Friedrich von Hardenberg undertook a two-week journey on foot through the area south of Leipzig. It was the last summer of his life. He was 28 years old, and had another nine months to live. He had already published a handful of poems, essays and a collection of aphorisms under the name 'Novalis'. Like Alexander von Humboldt, he had studied under the renowned geologist Gottlob Abraham Werner at the Mining Academy in Freiberg. Now he was involved in a major project of Werner's. The deposits of 'combustible fossils' in Saxony were to be located, described and mapped for the Mines Inspectorate in Freiberg. This substantial project was undertaken simultaneously with the 'Inventory and Regulation' of the Saxon forests which Heinrich Cotta was preparing in Tharandt. Novalis, who had just completed the first part of his novel *Heinrich von Ofterdingen*, hiked for two weeks – often cross-country – together with a fellow student from the mining academy, through the hilly landscape to the right of the White Elster river, right down to the area around Gera and Ronneburg. "On a summer morning I grew young . . ." So runs the great poem 'Astralis', probably written in these days, which was meant to open the second, never completed part of the novel; ". . . the world lay blooming round the bright hill . . ." Novalis sent his report to Werner in Freiberg. It is one of his last official acts in office. In August 1800 "expectoration of blood" sets in. In March 1801 Novalis dies in his parents' house in Weissenfels.

Werner's final report on the "geognostic investigation of the territory", which was based in part on Novalis's research, marks the beginning of the extraction of the central German lignite deposits. 200 years later, the strip mining here will leave behind a huge moonscape. What Novalis could never have imagined is that at the very end of his geological journey, all around Ronneburg, there lay beneath his feet huge deposits of uranium ore. Here – and in the nearby Ore Mountains – is where the Soviet Union mined a major part of its uranium supplies. Today, the landscape is being 'restored to nature'. The craters of the open-pit mines are mutating into an idyllic chain of lakes, the domed slagheaps from the uranium mining into green hills.

Novalis's short working life as a saltworks inspector was devoted in its entirety to exploring the lignite deposits. "The increasing wood shortage forced these people to give consideration to this cheap surrogate for wood, and now its use grows more widespread with every year," he wrote to Werner in 1799. The immediate aim of his investigations was to promote the burning of bituminous rocks, of peat and above all of lignite or brown coal, and wherever possible to prevent the use of wood entirely. Novalis calculated how much coal would suffice to replace a cord of wood when boiling salt, and tested which sort of coal was best.

In December 1799, however, his thoughts took a different direction. He wrote in a report to the finance ministry in Dresden that the use of wood, as well as that of coal, might be made unnecessary in a few years anyway. The production of salt by means of sunlight could replace salt-boiling with wood or coal. What Novalis was referring to here is an age-old process used already by the Incas and still practised today on the beaches of the Indian Ocean and elsewhere: filling shallow boxes or basins with concentrated brine and allowing the power of sunlight alone to evaporate the water, leaving behind pure sun-salt. In Novalis's Romantic economy, the future belonged to solar power.

A ticking time-bomb

At the end of the 18th century, just as the first Prussian coal mines were opened, a Berlin chemist set the clock ticking on a time-bomb which has yet to be defused. Martin Heinrich Klaproth, who had trained as a pharmacist, was teaching chemistry at the Prussian artillery academy in Berlin. Novalis had a textbook written by Klaproth on his bookshelf. Alexander von Humboldt knew him personally. Before Humboldt departed on his exploratory journey to the Bayreuth mines, he had several conversations with him to seek his advice. At that time Klaproth was already one of only eight natural scientists in the Class for Experimental Philosophy of the Prussian Academy of the Sciences which Humboldt was invited to join in the summer of 1800.

Klaproth carried out mineralogical research in his private laboratory. In 1789 he received – probably through an arrangement with the Mines Inspectorate in Freiberg – several samples of the mineral pitchblende. They came from mines in Johanngeorgenstadt and St Joachimsthal. Klaproth dissolved the pitchblende in nitric acid. He was then able to precipitate a yellow compound by neutralising the solution with sodium hydroxide. He mixed this compound with linseed oil and then burnt off the oil. The black powder which resulted was subjected to 'very severe heat' in a porcelain oven. By this means he obtained a 'grey, compact, hard and porous mass'. He named it 'uranium'. On that day, the 24th of September 1789 – two months after the attack on the Bastille in Paris – the uranium era, the nuclear era, could be said to have begun. The new element was soon enthusiastically taken up in the Ore Mountains region in Bohemia and

Saxony, first as a means of colouring glass, and then even for medicinal purposes. At that time, of course, nobody had any idea of the frightening energies slumbering within it. Five generations later it would look very different.

The solar era

"Will the current over-exploitation of coal . . . lead sooner or later to this precious natural power source running out completely? Or will we be able to put a stop to this immense waste of valuable national assets by exploiting the Sun or the self-renewing power of water?" These questions were posed in a leading article in the *Berliner Tageblatt* newspaper in the summer of 1930 to mark the opening of a global conference on the future of energy. "The Sun is the ideal source of energy." This contribution in the same edition comes from the venerable chemist and natural scientist Wilhelm Ostwald, a Nobel laureate. The conference brings a warning from him: "Fossil coal is a unique and unrepeatable inheritance which has fallen to us, and it is finite." But all energy sources derive ultimately from the radiant energy of the Sun. Only the rays of the Sun keep the wheel of life in motion. As a young man Ostwald had followed the debate over the laws of thermodynamics and the principle of entropy which derives from them. You can't have your cake and eat it, and pass it on to your children as well. Available matter transmutes into unavailable matter. Its finite nature is incontrovertible. So one has to ask whether sunlight could not be *directly* converted into electrical energy. Only by this means will mankind have unlimited access to the free energy necessary. At this point in his chain of thought, which he had set out in his 1911 book *The Mill of Life*, Ostwald brings in the "photoelectric current". Today we would call this photovoltaics. "If one places two lightly oxidised copper plates in specific solutions and connects them with a wire, then a current will flow along that wire when one of the two plates is illuminated." Of course, the currents observed up until that time had been far too weak for technical applications; but in principle, the direct generation of electricity from sunlight was possible. At any rate, he concluded, one could look forward with "equanimity and optimism" to the future for the coming generations, for our children and grandchildren. "As long as the Sun shines, they will not want for energy."

On the second day of the conference, the stage in the Kroll Opera House was taken by a man who had researched into the nature of light, and therefore of photovoltaics: Albert Einstein. A charismatic and fluent speaker, he soon has the audience of 4,000 under his spell. His demanding topic is the problem of space in physics, and his contribution to this field: the general theory of relativity. The Nobel Prize, however (for which Wilhelm Ostwald was the first to nominate him in 1910), was awarded to Einstein in 1922 not for his work on the theory of relativity. What won it for him was a discovery which he himself regarded as the

135

only 'really revolutionary' one of his career: the law of the 'photoelectric effect'. His discovery here represented a radical break with existing ideas. Light, he argued, was nothing but the flow of tiny packets of energy. "The energy of a light ray spreading out from a point source . . .", Einstein wrote in 1905, "consists of a finite number of energy quanta which are localised at points in space, which move without dividing, and which can only be produced and absorbed as complete units." If one of these quanta of light – a photon – hits a solid body at great speed, it is absorbed by an electron, and imparts to this the energy to free itself from the atom. This is how under specific conditions an electrical current is generated. What Einstein explained in fundamental terms in this theory of light is the photovoltaic effect, already well-known at that time. It was first observed and described by the French physicist Edmond Becquerel in 1839. The technical application of these discoveries in physics, however, via the development of a functioning solar cell, was then still only a distant possibility.

"Brighter than a thousand suns . . ."

The ultimate experience of 'shock and awe' – the sight of an atomic explosion – did not happen until 1945, in the desert of New Mexico. For the participants of the 1930 energy conference it is still unimaginable. But perhaps a premonition is in the air when the British astronomer Arthur Eddington, a friend and colleague of Einstein's, presents in the main hall a contrast to the gentle vision of solar energy. His talk ushers in the race of the technologies of the future. One day, he declares, when the finite energy sources have all been used up, we will use the energy contained inside the atom. For the life-giving energy from the Sun is almost certainly released by sub-atomic explosions taking place there. Infinite quantities of energy are therefore available from atoms. "We are not yet that far," Eddington says at the end of his talk. And if one day we finally do get there, he adds with a pinch of black English humour, he would rather not be present in the laboratory.

In the year 1930, there are a handful of people who do have the required sang-froid to make the attempt. If Eddington, during a break in the conference, had strolled out through the Brandenburg Gate and down 'Unter den Linden' to the main building of the Friedrich-Wilhelm University, he would have been able to pick up the lecture timetable, fresh from the printers. The winter semester of 1930/31 in the Physics faculty promises intellectual fireworks. The masterminds of the new physics are all there. The Nobel Prize winners Max Planck and Walther Nernst; the future Prize winner Erwin Schrödinger. Further down the list appear the names of the new young Turks: Lise Meitner and Leo Szilard will together lead seminars on "questions of atomic physics and atomic chemistry", and daily experimental classes on "the physics of radioactivity". If one adds in Eugen Wigner's seminar on "the structure of the atom and quantum theory" at the Technical University in Charlottenburg, then it does not seem absurd to think that in Berlin the splitting of the atom is not so far off. And indeed, by 1938 Lise Meitner, together with her long-time colleague Otto Hahn, will have solved the mysteries of nuclear fission. Szilard, by now an emigrant, will take part in the first controlled nuclear chain reaction at the University of Chicago in 1942, having got the Manhattan Project under way in 1939 with Wigner and Einstein. And Albert Einstein in 1945 will greet the end product of that project, the mushroom cloud, with two syllables: "O weh!"

The discovery of photovoltaics

"It is precisely in regard to the efficient generation of energy that the new photoelectric cell strengthens the hope that a technically exploitable way of transforming the huge quantities of energy radiated to us every day by the Sun will be found." This far-reaching sentence was formulated in the preparatory notes for a lecture, written at a desk in the Berlin suburb of Dahlem in those same summer days of 1930. Bruno Lange, an expert in physical chemistry, will not deliver the lecture until the beginning of September, at the German Physicists' Congress in Königsberg (now Kaliningrad). Lange is 28 years old. He completed his PhD under Fritz Haber, the inventor of artificial fertiliser – and of mustard gas – in 1927. Now he has a post at the Kaiser Wilhelm Institute for Research into Silicates in Faradaystrasse, just a stone's throw from the laboratories where Hahn and Meitner are working on nuclear fission. His "new type of photo-electric cell", Lange explains in the lecture, consists of two metal electrodes with a layer of copper oxide in between, with a barrier layer around the whole. When it is illuminated, photoelectrons are produced on the copper oxide, and the current flows through the barrier to the lower electrode. At the same time the outer electrode receives a positive charge, and a surprisingly strong photoelectric effect results.

The phenomenon described in Dahlem was not new in principle. The New York inventor Charles Fitts had discovered "the electromotive effect of illuminated selenium" in 1884. He had covered a metal sheet with a thin layer of selenium, and covered this in turn with a fine gold leaf, and used this to produce a current. He had sent some of his plates to Werner Siemens in Berlin. The 'Prince of Technology' was interested in everything to do with electricity. Siemens was impressed. He demonstrated the American solar cell at the Prussian Academy of Sciences. "As we are witnessing here for the first time the direct transformation of light-energy into electric energy," this discovery is "of the greatest scientific importance." Bruno Lange quotes Siemens's comments. He also refers to the experiments which Walter Schottky is carrying out, using similar processes, in the laboratories at Siemensstadt. Schottky and his colleagues had in fact already demonstrated, to public amazement, a small motor driven by solar energy. But the problem was that only a very small current could be produced. For the cautious man from Siemens, photovoltaics was still a toy. "Unfortunately, the gulf between what is possible in principle and what is technically possible is in this case, as in many others, much too big."

What neither he nor Lange had an inkling of was that a process which would give a huge impetus to photovoltaics at the end of the century was already lying unnoticed in a desk drawer. In the middle of the First World War, the Polish chemist Jan Czochralski, an employee at AEG who had studied at the Technical University at Charlottenburg and then become the head of the laboratory at 'Metallbank' in Berlin, had developed a process for making monocrystalline materials. The discovery had come to him by accident. He had dipped his pen into a crucible containing liquid tin instead of into the inkpot, and a fine thread had been formed. But his process for growing crystals was forgotten. In 1929 he accepted an invitation to join the University of Warsaw and died in 1953 in the town of his birth, Kcynia near Poznań. Around the same time his idea was rediscovered in the USA. Monocrystalline silicon became the material which enabled the development first of microelectronics and then of solar cells of an efficiency which had remained out of reach before.

It was clear to Bruno Lange that in 1930 the search for a new base material for photo-electric cells still had a long way to go. Nevertheless, he suggested in lectures and in interviews that the possibility of increasing the efficiency of photo-electric cells to a technically viable level was not so distant. This led, at the end of this remarkable year, to a media event which created unheard-of publicity for the technology of the future.

It was sparked by an article in the New Year edition of the *Berliner Zeitung* newspaper. Its author was Hans Dominik. Born in Zwickau, Dominik had studied mechanical engineering in Charlottenburg, had worked as an engineer and in advertising for Siemens & Halske around 1900, and then during the Weimar Republic had become a best-selling author of futuristic technological fantasies. Under the headline 'Wishful Thinking at the Turn of the Year', Dominik presents an overview looking both backwards and forwards. Inspired by the activities of the rocket scientists around Hermann Oberth and the young prodigy Wernher von Braun, which had made a big splash in Berlin in 1930, Dominik prophesies "rocket flights through empty space to the Moon". But then he turns his attention to the solar cell and the year just past. "They have succeeded in finding a combination of copper and copper compounds which delivers electrical energy directly when illuminated." Picking up on utopian visions which have been around for a long time, he paints pictures for his readers of installations in the

Sahara desert which gather and store sunlight in the form of electricity and deliver it to the energy-hungry Europeans.

Dominik's article was carried around the world and inspired euphoric enthusiasm for solar energy in 1931. The British *Science News* carried the headline "Invention makes electricity directly from sunlight." However, as people heard how inefficient the photo-electric cells were, the wave of enthusiasm soon ebbed away. Bruno Lange's superiors at the Institute for Research into Silicates were displeased by what they regarded as frivolous publicity. Their displeasure led to Lange's dismissal. In 1933 he set up his own independent company, which still exists today. His photo-electric cell became very successful. Not, however, in the energy sector, but as a component of measuring and analytical instruments. Bruno Lange died in Berlin in 1969, the year of the Moon landing. Without photovoltaics it would not have been possible.

In 1954, three American researchers presented the prototype of their 'Solar Energy Converting Apparatus' in public. Calvin Fuller, Gerald Pearson and Daryl Chapin had developed the first silicon solar cell at the famous Bell laboratories in Murray Hill near New York. The new basic material made possible levels of efficiency many times higher than those achieved with earlier models. The leap into technical viability had been achieved. The visions of a solar age acquired renewed plausibility and momentum. The American engineer Buckminster Fuller used the metaphor 'spaceship Earth' at around this time. It flies through space at a distance of 150 million kilometres from the Sun. Just close enough for the Sun to keep us alive. Just far enough away for it not to incinerate us. We get all the energy we need for the maintenance and renewal of life, and more, from our Sun and from the gravitational force of the Moon. The huge quantities of energy delivered every day by wind, water, tides and direct solar radiation are more than enough to keep all life processes running.

CHAPTER ELEVEN

Translating 'sustainability'

With the onset of the fossil fuel age, one country after another switched over its economy. No longer were they dependent on firewood as a source of energy. The idea, the term and the associated vocabulary of sustainability all disappeared from the mainstream. They survived, and even flourished and developed, in a well-protected niche – forestry. Forestry continued to be important, because wood continued to be an important construction material. Sleepers for building railways, timbers for shoring up tunnels in the coal mines, newspaper – only in the industrial era did these become mass-produced goods. Through the international communication networks of the forestry community the German term established itself worldwide. In the 19th century *sustainability*, or *sustained yield* became the Holy Grail of the international science of forestry. Its route to the international arena went via many stations. One of the first was a democratically organised, environmentally conscious and multilingual country – Switzerland.

The path over the Alps

The country between Lake Constance and Lake Maggiore, Lake Geneva and the valley of the Inn has a rugged, tough and self-willed character. Since ancient times the Swiss have regarded their forests as a precious jewel. The ecological role they play in addition to the purely economic was more obvious here than elsewhere. As was the case everywhere, they served to provide the populace with energy and with timber for construction. But the Swiss forests also played a key role in water management. In the high mountain regions they protected the settlements and traffic routes against avalanches. They protected the Alps (up until the 18th century this word meant the high pastures) against erosion and the low-lying towns against floods.

"The blue horizon encloses a wreath of shining peaks/ where the last rays break upon a black forest." Albrecht von Haller's didactic poem of 1732, 'The Alps', describes the high mountain landscape. With the settlement of the alpine valleys in the early middle ages, the period of the forest clearances began. From the river valley floors, the narrow strips of arable land and smallholdings expanded ever further upwards onto the mountain slopes. And from just below the tree line downwards, stands were cleared to create pasture for livestock. Between these two zones a belt of trees, as wide as possible, was left standing, and not used for any forestry purposes at all. These protective forests, sometimes called 'barrier woodlands', were legally protected already in the Middle Ages. The remainder of the common woodland was also strictly regulated. It was called the *Allmende*, or commons (here meaning common woodland). Those entitled to use it were allowed to remove only a strictly limited number of trees for their own domestic use. A method of selected single-tree cutting was developed called *Plenterwald* (sometimes translated into English as 'Plenter forest'). Large clearfellings were prohibited. The aim was to maintain over the longer term a proper balance between grain farming, pasture, woods and wasteland. For it was not only the immediate benefits from the woods which were important to the human population, but also their key role in the natural economy. Maintaining the woods was a matter of survival. Necessity dictated here that the economy had to become ethical and ecology political. This is the classic matrix for sustainability. The spectre of a sudden, unexpected and devastating wood shortage nevertheless continued to haunt Switzerland. Against the background of a chronic fear of the deforestation of the mountains and of the 'natural disasters' which would follow, this spectre was taken very seriously.

In the early 18th century 'economic societies' sprang up in the towns of the Swiss Confederation. Concerned citizens discussed in these gatherings the economic problems currently affecting their communities. From 1713 onwards, a *Holzkammer* ('Chamber for wood') in Berne concerned itself with all matters affecting forestry.

The polymath Albrecht von Haller was a prominent figure of the Swiss – and European – Enlightenment. The 'immortal' Haller (Herder) was one of the most famous doctors in Europe, a mathematician and botanist, an early patron of Linnaeus, and a poet of some distinction. His poetry, alongside the writings of his countryman and contemporary Jean-Jacques Rousseau, made the mountains fashionable – they had previously been regarded as an inhospitable wilderness. Haller had studied medicine in Tübingen and Leiden, and in 1736, aged 28, he accepted an invitation to teach at the newly-founded University of Göttingen. One year later he almost succeeded in bringing Linnaeus, who was the same age, to Göttingen. The plan was for the Swede to take over the famous botanical gardens (still extant today) alongside a professorial chair. Although in the end

Linnaeus did not come, Haller himself remained for 17 years in Göttingen. Subsequently he took over the direction of the Berne saltworks. In this capacity he took a keen interest in the forests of the canton, in order to ensure, as he wrote in 1765, that "the coming generations in this, as we hope, eternally blooming country . . . will have available to them . . . the necessary supplies of wood."

The membership of the Economic Society of Berne, of which Haller became the president, studied closely the relevant European literature – Colbert's *Ordonnances*, John Evelyn's *Sylva*, Carlowitz's *Sylvicultura Oeconomica*, Duhamel du Monceau's writings on forest botany, Moser's *Forest Economy* and of course the works of Linnaeus. The idea of sustainable forestry was therefore as familiar to them as the concept of the *Oeconomia naturae*. However, the word *Nachhaltigkeit* ('sustainability') did not enter the vocabulary of the Swiss forestry community until after 1800.

The man who introduced it was Karl Kasthofer, head of the forestry administration for the canton of Berne. Kasthofer had acquired his professional training in Germany: at the Faculty of Cameralism at Heidelberg University; then in Göttingen with the renowned economist Johann Beckmann, a student of Linnaeus; then finally at forestry schools in the Harz Mountains. He knew the writings of Hartig and Cotta very well.

In 1818, in his *Observations concerning the Forests of the High Mountains of Bern*, Kasthofer addressed the proper relationship between consumption and reproduction of wood. He defines the term *nachhaltig* (in a footnote at this stage) as follows: "A forest is being used sustainably (*nachhaltig*), when each year no more wood is harvested than nature annually produces there, and also no less." He introduces it as "a technical term coined in Germany". In Kasthofer's later writings the term acquires a central role and value. However, his focus shifts ever further away from the sustainable 'use' of the woods and towards a sustainable 'yield', or more precisely the maximum monetary yield. It is with this emphasis that Kasthofer's concept of sustainability is translated into French.

The canton of Berne is bilingual. The French-speaking regions lie in the densely-wooded mountains of the Jura. The publications of the cantonal forestry administration are routinely translated into French. The anonymous translator renders Kasthofer's *nachhaltigen Ertrag* ('sustainable yield') as *produit soutenu et égal d'une forêt* – sustained and constant product of a forest. This is a significant semantic event: the word *soutenir* appears for the first time here in the vocabulary of sustainable forestry – an important step on the way to the coining of the term 'sustainable development' by the Brundtland Commission.

But compared to the broad German term *Nachhaltigkeit*, the concept of *produit soutenu* narrows the focus noticeably on the yield. The choice of words for the translation accurately reflects Kasthofer's interpretation of the term. The Bernese master forester is namely an adherent of Adam Smith's doctrine of the free market. He regards the prognosis of a coming wood shortage as a figment of the imagination. For him, the regulation of forestry in the cameralistic traditions of the German principalities is outdated. The rigid division of the forests into yearly felling plots, and the long-term economic planning, are appropriate to neither the topographical nor the political situation of Switzerland. He forcefully advocates the privatisation of the commons and of the publicly-owned woodlands (with the exception of the protective forests). He calls for the deregulation of the timber trade. The free play of supply and demand will ensure *Nachhalt* – here meaning a constant level of supply – as if by an invisible hand. If the yearly harvest in these woods is below its sustainable level, then the price of wood will be higher than its real value; and if the harvest is greater than the sustainable level, then the price of the wood will be lower than its real value. Kasthofer's liberal line is not shared by the patrician governing body of Berne. Because of his deep knowledge of forestry he is not dismissed from his office. But his successors reverse his efforts to liberalise the forestry management of the region.

Their criticism is sharply worded. "The more thoroughly one tests this so-called theory of balance, under which it is proposed that the power to destroy, to create and to maintain the forests should depend solely and unrestrictedly on the price of wood, the more certainly must one conclude that this theory is nothing more than a dream. . . . Whoever can speak in such terms knows neither people nor trees, nor by extension the influence of these last on the economy of nature."

A critique of the all-powerful free market, in the name of the natural economy – of ecology. It was formulated in 1849 by the Bernese forester Xavier Marchand. Like Kasthofer, he had received his training in Germany, namely in Freiburg and Munich. His departmental colleague and supporter Elias Landolt was a graduate of the Forestry Academy at Tharandt. Landolt began his career studying the high mountain forests and watersheds. Like Marchand in Berne he stressed the protective function of the Swiss woods and thus the ecological component of sustainability. In 1855 the Federal Polytechnic Institute was founded in Zürich, known today as the ETH (Eidgenössische Technische Hochschule, or Swiss Federal Institute of Technology) Zürich. Landolt took on the directorship of the forestry faculty. In the 20th century the ETH became one of the most renowned universities in the world, and Landolt's faculty one of the most famous institutes of forestry, its hallmark being close-to-nature silviculture.

And today? Switzerland was the first country in the world to embed the concept of 'sustainability' in its constitution. "The Swiss Confederation", so runs Article 2, "shall promote the common welfare, sustainable development, internal cohesion and cultural diversity of the country." This provision has been in force since 1st January 2000.

Tharandt and Nancy

After the French Revolution, the authority of the all-powerful Maîtres des Eaux et Forêts collapsed. Parts of the royal domains, of course, were taken over and maintained as public forests. In some areas they were even expanded through the incorporation of woodland expropriated from the clergy. But the principles of *bon ménage* were now regarded as a cunning trick perpetrated on the Third Estate and the lower classes by the Ancien Régime in order to keep them out of the woods and to protect the privileges of the old elite. Now the ardent and ubiquitous cry was for *égalité*. This cry led all too often, of course, to a demand for privatisation. But would social justice be promoted by *laisser faire* in the forests? With the Restoration, the wind turned once again. The complete liberalisation of access to the woods was no longer considered to be in the national interest. It was against this background that the advantages offered by the new German ideas of forestry began to gain recognition in France. So the intensive exchange between the two neighbouring countries which had been under way since Colbert's reforms acquired renewed currency. Now the idea of the *conservation des bois* was re-imported – as sustainability.

Although research into the botany of the forests had been raised to a new level under the Ancien Régime by naturalists like Buffon, Réaumur and Duhamel du Monceau, little use had been made of the new scientific knowledge in the forests themselves; rather, the simple bureaucratic campaign against illegal cutting had continued. But of course this was not sufficient in itself to restore them. What the French found particularly interesting in the new German concept of sustained yield forestry was the method of natural regeneration.

1805 saw the publication in Paris of a bilingual edition of Georg Ludwig Hartig's classic, *A Guide to Silviculture*. Other translations of German writings on forestry followed quickly. Around this time, the young French forester Bernard Lorentz became acquainted with German forestry through working for the Napoleonic administration in the Rhineland Palatinate. In 1815, the Alsatian Adolphe Parade went to study under Heinrich Cotta at the newly-established Royal Saxon Forestry Academy at Tharandt. After qualifying, with distinction, he hiked for several months through the German forests in order to study at first hand the practice of regulated forestry.

In 1821 the government in Paris passed a new 'Code forestier'. This was the period when a system of 'Écoles supérieures' was being established for the education of a new technocratic elite, and amongst these was an institute for training the next generation of the Republic's foresters. Nancy was chosen as the location. This was no ancient Romantic town of crooked alleyways hidden in a remote wooded valley, like Tharandt, 600 kilometres to the east. With its strictly symmetrical

Baroque streetplan, its regular town squares, palace, triumphal arch, water fountains and parks, the former capital of Lorraine was – and is – one of the most
resplendent of French provincial cities.

As founding Director of the new forestry academy the government
appointed Bernard Lorentz. He in turn made Adolphe Parade his assistant and
eventual successor. The curriculum was shaped by the new German forestry
ideas. German was a compulsory subject. In the library, the classics of German
forestry stood side by side with Colbert's *Ordonnances* and the writings of
Buffon and his followers. The works which Lorentz and Parade produced themselves dealt with *réensemencement naturelle*, or natural regeneration, and with
methods of measuring forests and trees. Instead of *conservation des bois*, the new
guiding idea is now *principe du rendement soutenu* – sustainable yield. The formation of the terminology in France was similar to that in western Switzerland.
It focused on the sustainability of the yield.

Traditional French forestry was modernised with the aid of the new German
thinking. However, numerous foresters and forest owners in this broad country
between the Ardennes and the Pyrenees criticised the rigid, schematic approach
and the shift to monocultures. They mourned the loss of the magnificent woods
of *la douce France*. Nevertheless, Nancy flourished. Like Tharandt, in the course
of the 19th century the 'École' became famous around the world.

Our modern concept of 'sustainable development' represented a problem for
French national language policy. Presumably in order to get away from the English precedents in the Brundtland Report and Agenda 21, the translation chosen
at first was *développement durable*. That was not a long way off, as *durable* means
enduring, steady, stable, lasting. Meanwhile, *développement soutenu* has also
established itself in French.

In the forests of the north

In the summer of 1858, Baron Edmund von Berg is journeying through Finland.
At this time, he is the Director of the Academy at Tharandt, a successor of Heinrich Cotta. He learned his trade, and his scientific knowledge, in his home town
of Göttingen and in the forests of the Upper Harz mountains. Berg is a famous
hunter, an expert on the burning of charcoal, and a critic of coniferous monoculture forests and of the liberal 'soil rent theory'. A portrait of the time shows him
as a broad-shouldered, corpulent, full-bearded forester, a worthy representative
of the famous Tharandt school.

He has come to Finland at the invitation of the Imperial Senate. Since 1808,
the country has been an autonomous Grand Duchy of the Russian Tsarist empire.
Its official language is still Swedish. The man who governs here in the name of

the Tsar is a namesake of the German forester: Duke Lambert von Berg is the scion of a German-speaking Baltic noble family, a Tsarist general and veteran of the campaigns against Napoleon, and a seasoned administrator of Tsarist colonialist policy. He wants to make Finland into a model Russian province. Sawmills right across the country will open up the huge timber stocks of the forests for commercial exploitation. For this ambitious and 'progressive' project, the condition and quality of the Finnish forests is to be assessed by a leading European expert.

For the duration of the short Nordic summer, by horse and coach, sometimes by boat, Edmund von Berg and his small team travel ceaselessly around this huge, sparsely populated country: from Helsingfors, now Helsinki, up to Rovaniemi on the Arctic Circle, and further northwards on the Ounasjoki River, where the woods and moors gradually give way to treeless tundra – the wilderness of wolves, beavers and reindeer herds. At the northernmost point of his journey, his gaze sweeps over the last utilisable timber of any value. He advises the local dignitaries not to set the rotation period at less than 250 years. So in August 1858, deep inside the harshness of the Arctic Circle, this man from the idyllic setting of Tharandt is planning the forests of the 22nd century. Von Berg returns to the capital on a wide sweeping route which takes in a sortie to Lake Ladoga. Finland was then, as it is now, the country with the most extensive forest cover in Europe. More than three quarters of its territory is covered with *tall*, *gran* and *björk* – pine, spruce and birch. Where, if not here, could wood supplies be described as inexhaustible?

Berg's verdict is sobering. He refers to the flames and smoke from the wood fires which he and his companions saw around him almost daily, sometimes dangerously close. In the province of Österbotten, the production of tar for the European shipbuilding industry has already devoured vast stocks of trees. On no account must the new sawmills be allowed to continue the plunder. You Finns, von Berg tells his hosts, are like the woman in the fable who in her ignorance and greed kills the goose that lays the golden eggs. "Sad" and "depressing" – this is how he summarises the effect on him of seeing the "plundered, devastated and burnt-out forests of Finland".

Progress? von Berg believes that progress is being used here to cover up an unthinking business as usual. In Finland, he says, a state has already been reached "which, if it is regarded as 'progressive', must give rise to the gravest concern". Truly progressive thinking, in his view, must take account of the complex ecological factors. If things are allowed to progress in the same "thoughtless" way as hitherto, then the worst possible devastation of the forests must be expected and anticipated. The deforestation will inevitably bring about a deterioration in the climate in the affected areas and will spread out over huge areas "like a cancer". For forests always serve to regulate climatic conditions. Without them the climate in the far north would be harsher and less hospitable.

What to do to ensure that the land remains habitable? Von Berg's answer is simple: sustainability. In his Memorandum for the Imperial Senate of August 1858 he sets out the basic principle: introducing management methods which ensure a constant and steady yield in the future. "[F]or the living and the coming generations have an equal claim and equal rights." And therefore one has to act not only with an eye to the future harvest of wood but at the same time bearing in mind the regeneration or the protection of what we have.

At that point the Chancellery of the Finnish Senate had a problem: how to translate this word used in von Berg's report? There was no Swedish word for *Nachhaltigkeit*. Carlowitz's *Sylvicultura oeconomica* had been read in small Swedish-speaking circles. It was already cited in a paper of 1747 on tar production by a pastor from Österbotten. But nobody had ventured to translate Carlowitz's new term. The solution at first was to reach back to the Swedish form of the long familiar term *conservatio*. *Konserverande skogsbrok* – conserving, or sustaining forestry – and that means *med omsorg för skogens återväxt och bibehållende vid ständigt lika stor virkesmassa* – providing for the regrowth of the forests and maintaining a constant stock of wood. This was certainly precise, but perhaps a little long-winded. In the years which followed, sustainability was translated into Swedish simply as *uthålligheit* (durability). After Finnish was raised to the status of the second official language in 1863, the word *kestävyys* (literally, 'permanence') was also used. The basic ideas fed into the Finnish forestry law of 1886.

The idea and the term became a part of the general culture and of the perceived national interest in Finland and the other Nordic countries. Certainly, Olof Palme and Gro Harlem Brundtland were drawing on support from this tradition during the 1980s when they fought to put sustainable development – *hållbar utveckling* – on the global agenda. When international comparisons are made today of the implementation of sustainability objectives, Finland and the Scandinavian countries always appear towards the very top of the lists.

'Sustained yield' in the jungle

In *The Jungle Book*, Rudyard Kipling's entertaining fantasy story, the camp fires of the woodcutters burn only a leap away from the wilderness where Akela's wolf pack, the black panther Bagheera, the python Kaa and all the other animals of the tropical rainforest defend their habitats among the vines, orchids, bamboo canes and giant trees. On the Indian subcontinent there are 1,200 native tree varieties, nearly ten times as many as in Europe. Is forestry on a European model conceivable in such tropical conditions?

When Rudyard Kipling had just celebrated his first birthday, growing up – unlike Mowgli – safe and secure in Bombay, a German forester embarked on a bold plan to transfer the principles of regulated silviculture to the Indian jungle. In 1864 the British colonial government appointed Dietrich Brandis to the post of Inspector General of Forests in India. The man who was to spend the next 20 years designing management plans for woodlands scattered over the vast territory stretching from the Himalayas to Ceylon (Sri Lanka) and from the Hindu Kush to the Gulf of Siam came from Bonn. His background was in botany. He began his career as the keeper of the botanical gardens of his home town. His knowledge of scientific silviculture was largely self-taught. He first implemented it in the rainforests of Burma. The teak trees which fetched the top market prices only accounted for 10% of the forest there. The 'Brandis system' was intended to transform the forest into a teak monoculture. Just as in Tharandt, timber stocks and growth rates were calculated, rotation periods established and management plans drawn up. As the new head of the colonial forest administration for the whole of India, Brandis drew a number of strategic threads together. He continued to set up plantations for teak and rubber trees, using the strategy of sowing and planting typical of British forestry since John Evelyn. He introduced the Tharandt method of forest management planning and transformed richly diverse forest habitats into monocultures of the desired tree species. And he created protective belts of woodland which were not exploited at all in order to safeguard the water cycle and the climate. On the basis of this programme Brandis built up an effective forestry department. All British trainees were sent off to be educated either at one of the German forestry academies or at Nancy.

When he retired in 1881 and returned to Germany as Sir Dietrich Brandis (having been knighted for services to the British Empire), he was succeeded by his student and colleague Wilhelm Schlich, also a German. Schlich was born in Darmstadt, and had studied in Giessen, where the writings of Georg Ludwig Hartig were still compulsory. Alongside his onerous duties in India he took on two far-reaching projects. He organised the establishment and development of an institute of forestry for the British Empire – the Forestry School at Coopers Hill, which was incorporated into the University of Oxford in 1905. And he wrote (under the name William Schlich, which he adopted after becoming a British citizen) his monumental *Schlich's Manual of Forestry*, the first volume of which appeared in 1889. This was to remain for many decades the standard reference work in the English-speaking world.

Schlich sets out his credo at the very beginning. "Forests are, in the economy of Man and nature, of direct and indirect utility, the former through their produce, and the latter through the influence which they exercise upon climate, the regulation of moisture, the stability of the soil, and their sanitary, ethic and aesthetic effect upon Man."[1] And so it continues. All five volumes of the work are imbued with the spirit of British utilitarianism, with the terminology of German forestry and with the aim of combining ecological perspectives with practical economic interests. A forest namely offers not least "a convenient opportunity for the investment of capital and the profitable utilisation of inferior land". However, Schlich was well aware of the basic idea of sustained-yield forestry, that is – in his words – "to limit the extraction of produce to what the forest is capable of producing". In the volume on forest management, Schlich defines as the paramount objective of the forester's work "to give a sustained yield of produce in the future". A management based on the principle of sustained yield – this is Schlich's guiding principle for the forests and jungles of the Empire.

Here we have then the English version of sustainability. Like the French formulation *rendement soutenu*, 'sustained yield' is based on a derivation from the Latin *sustenire*. The term puts the emphasis on a constant maximised yield from the forest and requires rational forest management. With this emphasis, and mediated through Brandis and Schlich, sustained yield forestry arrived in the USA.

The American way

Around 1900, the timber industry was the third largest industry in America. The powerful timber companies were extremely profitable, but also incredibly destructive and wasteful. Silviculture simply did not exist. Not a single hectare of forest in the USA was subject to any kind of forestry management. What would have been the point? The forests were regarded as inexhaustible, and at the same time as an obstacle to the opening up of the country to commercial and industrial development. To be sure, there were the two National Parks, the first in the world: Yellowstone, established in 1872, and Yosemite, in 1890. But outside these protected zones the plundering of the country's natural assets went on unchecked. 'Cut out and get out' – clearcutting virgin forests and moving on to repeat the exercise elsewhere – was the infamous motto of the timber barons.

This first began to change only around 1900. Gifford Pinchot became the most influential administrator of forestry in the USA. He was from a family of French descent which had made its substantial fortune from Pennsylvanian timber. In 1889 he went to Europe to study forestry. Dietrich Brandis, who at that time was living in retirement in his home town of Bonn, became his mentor and lifelong idol. Brandis encouraged his young American visitor to spend a year at the French 'École nationale forestière' in Nancy. In 1890, he took the young man with him on an extended tour of the forests of central Europe which took in the oak forests of the Spessart region, Tharandt and the Black Forest. A trip to the Vienna Woods and practical training in the Bingen Forest completed Pinchot's education.

A particularly strong impression was made by a stay in Switzerland. In Zürich he talked to Elias Landolt, the Swiss advocate of close-to-nature forestry, and spent a couple of weeks as a trainee in the Sihlwald, Zürich's communal forest, a beech wood that had been managed sustainably for a thousand years. Under the guidance of forester Ulrich Meister he received at first hand a lesson in the Swiss version of *Nachhaltigkeit*. The German version, which regulated everything down to the minutest detail, had not corresponded with his own vision of silviculture . The Swiss understanding of ecologically sound forestry was much more to his liking. What particularly struck the young American during his stay in Zürich was that the municipal foresters were democratically elected and felt a deep responsibility towards the wider local community.

The guiding principle which Pinchot followed in building up public forestry in the USA after his return was 'wise use'.[2] It owed much to both *bon usage* (Colbert) and 'sustained yield' – the German concept of sustainability. Its philosophical background was the ethical doctrine of utilitarianism, whose classical definition – the greatest good for the greatest number – Pinchot extended to include the dimension of time. He defined wise use as "the use of natural resources for the greatest good of the greatest number for the longest time."

'Wise use' became the motto of the Conservation movement. 'Conservation' here means a usage which cares for and maintains the full potential of the forest to regenerate, rather than simply the protection of nature. Prohibiting the utilisation of natural spaces was known in the USA as preservation. This kind of strict protection of nature was applied only in the territories of the National Parks. But for all its utilitarianism, Pinchot's approach was a clear rejection of the cut-out-and-get-out methods of the American logging industry. The timber barons at first vehemently opposed the new approach. Then they swung around and adopted 'wise use' into their own vocabulary, albeit without making any substantial changes to their actual methods. An early example of 'greenwashing'.

In the 1930s, in the southern plains of the United States, "the earth ran amok", to use a famous expression of the American historian Donald Worster.[3] The so-called 'Dust Bowl' coincided with the Wall Street Crash of 1929. The Great Depression which followed led to hitherto unknown mass unemployment and poverty. The collapse combined ecological, economic and social dimensions. The Dust Bowl was to become the most severe drought in the known climate history of the USA. The land of the Great Plains between the Mississippi and the Rocky Mountains dried out and broke up. The water table sank. Dust storms of unprecedented ferocity sprang up and darkened the skies as far as Chicago. The Dust Bowl lasted for a whole decade. Every spring, the earth ran amok again. In the 'Bible belt' of the southern states many of those affected believed it to be a punishment from God. Others sought the causes in industrialisation and in the 'Fordist' methods increasingly adopted in agriculture. The First World War had driven wheat prices sky-high. Every part of the prairie was turned over to the plough. The farms were transformed into agricultural factories. A phalanx of combine harvesters and soil-tilling machines marched across the land. At the beginning of every growing season, the topsoil over vast areas was exposed, unprotected, to the wind, and was blown away.

The nation reacted quickly and decisively. The New Deal, President Franklin D. Roosevelt's political strategy against the economic and social crisis, was given a strong ecological component. Conservation became one of the pillars of the new policy. The newly-established Civilian Conservation Corps, a job creation scheme on a vast scale, set out on a gigantic programme of reforestation and renaturation of the landscape.

One of the experts who worked on the plans for ecological transformation was Aldo Leopold. Born in 1887 in the Midwest as the son of German immigrants, he had studied at the Yale School of Forestry founded by Pinchot's family, and had started out on his career as a forester under Gifford Pinchot himself. As a specialist in the biology of game in the Southwest of the USA, he developed an early interest in the new science of ecology. He believed that in the interests of their own survival, people had to learn to think in overarching ecological connections. "Think like a mountain" was Aldo Leopold's motto.

At the time of the Dust Bowl he was working in Wisconsin on programmes for the maintenance and restoration of land health, that is, of ecologically intact landscapes. He was critical of the fact that the patterns of thought which had led into the crisis were continuing unchallenged. The soil should not be viewed as a food factory. "The real end (of conservation) is a universal symbiosis with land, economic and aesthetic, public and private," he wrote in 1933. How is it possible that Congress cares about protecting migratory birds at a time when fellow-humans lacked bread, a senator from Michigan had furiously asked. "The trouble is," Leopold calmly answered, "we can never be sure which is cause and which is effect." And he attacked head-on the technocratic utopianism of "salvation by machinery", and the great illusion that "if we all keep warm and full, and all own a Ford and a radio, the good life will follow".

In the era of the New Deal, the German concept of sustainability suddenly provoked interest. In 1934, President Roosevelt, despite his deep aversion to the Nazi regime, sent an American forestry delegation to Germany. In Tharandt they received a notably warm welcome. In the land of his ancestors Aldo Leopold, who was a member of the delegation, maintained a detached perspective. He paid due acknowledgement to German legislation for the protection of nature. But he referred in gently disapproving terms to the Cubism of German forestry.

"That epidemic of geometry which blighted the German mind in the 1800s" had turned the forests into "wood-factories". The rivers were imprisoned in "strait-jackets" and reminded him of "dead snakes". The extermination of predators like the wolf, the lynx and the eagle had left behind "an eerie silence". Was that sustainable? Leopold thought that the suppression of everything that was inde-pendent in nature and its management by human hand would lead to the loss of ecological stability and thus of sustainability. The only enduring positive impres-sion he took away with him came from an excursion to the woods owned by Arnold von Vietinghoff-Riesch, a landowner and lecturer at Tharandt. His con-cept of the 'permanent forest' placed the potential natural vegetation and the continuity of the organism of the woods at the centre of the practice of forestry.

In the years which followed, Aldo Leopold expanded his ideas into a Land Ethic. "Quit thinking about decent land-use as solely an economic problem. Examine each question in terms of what is ethically and esthetically right, as well as what is economically expedient. A thing is right when it tends to preserve the integrity, stability, and beauty of the biotic community. It is wrong when it tends otherwise."[4] Aldo Leopold died in 1949. He was one of the very first thinkers and writers worldwide who combined the traditional terminology of sustained-yield forestry with the vocabulary of scientific ecology. Influenced by the British ecologist Charles Sutherland Elton, he used that term as early as 1935. His Land Ethic only became well-known around the world through the environmental movements of the 1970s. His innovation – incorporating ecology among the cen-tral precepts of sustainability – opened doors for others who came after him.

Gifford Pinchot survived Leopold by one year. His struggle had not been in vain either. In 1960, the principle of sustained yield became law in the USA. After decades of campaigning, the Multiple Use – Sustained Yield Act was passed. It defined sustained yield as "the achievement and maintenance in perpetuity of a high-level annual or periodic output of the various renewable resources without impairment of the productivity of the land."

Breakthrough

In 1951 the Food and Agriculture Organisation (FAO), the global nutrition organisation of the UN, was debating its 'Principles of Forest Policy'. The for-estry department was led by two men who together represented the traditions of European forestry. Marcel Leloup had been head of the venerable administra-tion of the 'Eaux et Forêts'. His deputy was Egon Glesinger, whose parents owned huge tracts of forest near the little town of Teschen (or Cieszyn, or Český Těšín) in the foothills of the Czech Beskids. He had had to flee Europe at the outbreak of World War II because of his Jewish ancestry. As an emigrant in

America Glesinger wrote a visionary book entitled *The Coming Age of Wood*. It was not oil but wood, he argued, which would determine the future of humanity. These two foresters fused the European traditions with the American approach and firmly anchored them into the principles of the FAO. "Subject to wise conservation and utilisation (forests) constitute an indefinitely renewable source of products which are indispensable to man." Each country therefore should seek "to derive in perpetuity, for the greatest number of its people, the maximum benefits available from the protective, productive and accessory values of its forests." Forests should be managed in such a way as to obtain "a sustained yield" (*rendement soutenu*). Wise use, sustained yield, in perpetuity: the idea and the concept of sustained-yield forestry, developed over centuries from the forest reforms of Renaissance Venice and Carlowitz to Pinchot – now it had arrived in a major think tank of the UN. The spread of the concept around the globe accelerated the thinking and discussion about what it meant. New, topical definitions emerged.

In 1968 William A. Duerr, a leading American forester and an expert on German forestry, defined the doctrine of sustained yield as follows. "To fulfil our obligations to our descendants and to stabilise our communities, each generation should sustain its resources at a high level and hand them along undiminished. The sustained yield of timber is an aspect of Man's most fundamental need: to sustain life itself." [5] This modern definition of sustainable forestry contains in full the essence of the Brundtland formulation. Like a pearl in an oyster. However, extracting the idea from its small, hard shell, and embedding it in a new, far broader context required another 20 years.

Around 1968 the era of 'Earth politics' began. The driving force behind this development was a strong feeling that a turning point had been reached on Spaceship Earth. Several long waves of historical development were coming together. Colonialism had come to an end, politically and morally; life together in the global village was now guided by other and fairer rules. The fossil fuel economy was no longer a part of the solution but a part of the problem: the exhaustion of the deposits was now in the foreseeable future; the pollution of the environment had become insupportable; the technical alternatives had matured and become attractive. Finally, an awareness had grown that the old individual and collective strategies of self-preservation didn't make sense any more. The war against nature, and the *bellum omnium*, the war of all against all, were coming to be seen as philosophical dinosaurs. The way seemed to be opening up for a peaceful civilisation in harmony with nature. However, at each juncture of its development, the new way of thinking bumped up against the hard constraints of power politics and of international diplomacy. To establish and implement principles like ecology and sustainability required extraordinary stamina.

What the German economist Ernst Ulrich von Weizsäcker had called 'Earth politics' came to be known in the 1990s by the academic term 'global governance'. *Gubernatio* – steering the direction of the world – had come back into the vocabulary of humankind. Not, as it had been in the Middle Ages in Europe, as an instrument of Divine Providence, but as the task of the United Nations, of national governments and of global civil society.

CHAPTER TWELVE

Earth politics I: new beginnings

Limits to growth

> We are searching for a model output that represents a world system that is: 1. sustainable without sudden and uncontrollable collapse; and 2. capable of satisfying the basic material requirements of all of its people.[1]

The word *sustainable* appears for the first time in its modern, broader meaning in 1972 in the Report to the Club of Rome on 'The Limits to Growth'. The word seems at first sight unremarkable. However, it already has a key function in the text. The report addresses the search for a world system which is capable of supporting human life. Once again, as in the Renaissance and in the early Enlightenment, what is at issue is the self-preservation, indeed the survival of the human race and the habitability of the planet. Once again it is all about confronting the real threat of collapse.

In April 1968, a small group of experts drawn from around the world gathered in Rome. It included scientists, science administrators and representatives from the OECD and UNESCO. The select group met in noble surroundings. The Villa Farnesina is an ornate Renaissance building with a secluded garden on the banks of the Tiber in the heart of Rome. The rooms are decorated with frescoes. One of the ceiling paintings depicts, in luminous colours, the night sky of the 29th November 1466. Cardinal Nicholas of Cusa had died only a couple of years before that day. A new policy for the forests was being planned in Venice. Copernicus had not yet been born. What the initiators of 'Project 1968' were contemplating was nothing less than a new 'Copernican revolution'. This meeting in April 1968 marked the birth of the Club of Rome.

It was held at the invitation of the Italian businessman Aurelio Peccei. Born in Turin in 1908, he had studied and completed his doctorate in the Mussolini era – his dissertation, amazingly enough, was on Lenin's New Economic Policy. In the 1930s he had worked for Fiat, helping to set up an aeroplane factory in China. Back in Italy, he joined the Resistenza, the anti-fascist resistance. In 1944 he was arrested and spent a year in prison. After the war he was sent by Fiat to work in Argentina. Next, he worked as a consultant to the ailing Olivetti business and then oversaw the construction of a car factory for Fiat in the Soviet Union. Peccei not only had an excellent understanding of technological and economic developments in Western industrialised societies but was equally familiar with the existential problems of the post-colonial 'third world'. Moreover he also had good contacts in what was then the Eastern bloc. His passion was 'global Earth' – *unus mundus*, as the ancient Romans had it. Peccei foresaw the end of the nation-state and a process of progressive globalisation, or as he preferred to call it, "planetisation". Peccei's view of the world was based on the conviction that there was an inescapable and irreducible interrelationship between the human economy and the biosphere.

But the consensus among Peccei and the others gathered in Rome in 1968 was that this interrelationship had entered into a deep crisis. In order to probe more deeply into what they called the 'world problématique', or the 'meta-system of problems', they arranged to meet for regular discussion rounds. The location was to be the Battelle Institut in Geneva, the city of Rousseau. This is how the Club of Rome came about. Its first big project was called Project on the Predicament of Mankind.

Among the founding members was the Austrian astrophysicist Erich Jantsch, who was employed by the OECD as a futurologist and was exploring aspects of the 'self-organisation of the universe'. He drew attention to the potential for their work of cybernetics and systems theory. In the autumn of 1968, at an OECD meeting on methods of future planning, the Club made its first contact with the systems theorist Jay W. Forrester from the world-famous Massachusetts Institute of Technology (M.I.T.) in Cambridge, USA. Forrester had started his career working in planning for the Allied armaments industries in the Second World War, where he had learnt two lessons: to think on a large scale, and at the same time to look for quick solutions. His toolkit comprised cybernetics and systems analysis, and his method was computer modelling.

In 1969 the German Eduard Pestel, a professor of Mechanics at the Technical University of Hanover, joined the Club of Rome. Only a few months later, in March 1970, Pestel outlined the Club's concerns in a presentation in Hanover to the top management team of the tyre company Continental. Pestel listed 35 problems which affected the whole human race and characterised all of them as 'component parts of a meta-problem'. He foregrounded five common variable factors: population growth, food production, industrialisation, resource stocks and environmental pollution. In order to map and describe the interconnections and links a 'world-model' was needed which would demonstrate the necessity for urgent action. Something had to be done, and it had to be done "on the basis of a set of values which would be far removed from the naïve faith in growth, the affluent complacency and the egotistical consumption" of contemporary society.

Three months later, members of the Club met at M.I.T. Jay W. Forrester had invited them to hear an introduction to the methods of system dynamics. The lead presenter was the 28-year-old Dennis L. Meadows. His idea was to depict the development to date of the five variables affecting economic growth, then to use computer modelling to extrapolate future developments from these data and to construct a 'world model' on that basis which would graphically draw attention to humanity's dilemma. His presentation convinced his audience. Meadows was asked to assemble a team for the project. The Club of Rome persuaded the Volkswagen Foundation to finance it: in November 1970 the German organisation made 775,000 Deutschmarks available. However, some scepticism was voiced. Some reviewers criticised the 'markedly technocratic character' of the project and the 'vague doom-laden catastrophe scenarios', which would enable only unspecific proposals to emerge. If the ecological problems were not laid at the door of the respective interest groups responsible, there was a danger that the only conclusion possible would be that of 'a general collective responsibility of all mankind'. These warnings accurately anticipated much of the later criticism.

The team began its work at M.I.T. at the end of 1970. It consisted of 17 people, whose average age was below 30. Twelve men, five women; ten Americans, three Germans, and one from each of Turkey, Iran, India and Norway. The group presented a summary of their final report to the Club of Rome one year later. The authors were named as Dennis L. Meadows, the systems theorist; his wife Donella Meadows, a bio-physicist, responsible for the language of the text; the young Norwegian environment expert Jorgen Randers; and the American oceanographer William W. Behrens. The report appeared as a book already in March 1972. Its title was *The Limits to Growth*.

> Suppose you own a pond on which a water lily is growing. The lily plant doubles in size each day. If the lily were allowed to grow unchecked, it would completely cover the pond in 30 days, choking off the other forms of life in the water. For a long time the lily plant seems small, and so you decide not to worry about cutting it back until it covers half the pond. On what day will that be? On the twenty-ninth day, of course. You have one day to save the pond.[2]

A children's riddle, well-known in France, and one with a striking answer. Retold in the first chapter of the Report, which deals with "the nature of exponential growth", it transports the reader to an idyllic setting – the feelgood biotope of a garden pond, whose centrepiece is a beautiful and fast-growing plant. But the idyll is deceptive. The growth of this one component is fatal for the whole. The end result is ecological collapse, which can only be prevented by quick and decisive intervention on the part of the gardener. For the water lily hides a dangerous secret. Its growth is not linear. That is, it does not grow each day by, say, one square metre. The surface area it has expanded to cover by any given day is always increased by a set percentage. It grows exponentially. And that, according to the analysis carried out by the research team around Dennis and Donella Meadows, is the problem with industrial society.

Following the specifications laid down by the Club of Rome, the group had focused its investigations and data collection on five main trends: accelerating industrialisation; rapid population growth; widespread malnutrition; depletion of non-renewable resources; deteriorating environment. In their computerised scenarios, they extrapolated those current trends in order to look into the future.

The worrying message from the computerised world-model was that there was exponential growth in all these fields. Positive feedback loops – vicious circles – between the individual fields further strengthened the dominant trajectory in each. Population growth increases both the cultivation of food and industrial production. The greater the growth of production and consumption, the faster the available resource stocks shrink, and the faster the decrease in the capacity of the Earth to provide nourishment, and in the capacity of the 'sinks' to absorb waste.

But one simple truth persists: the Earth is finite. This means that an upper limit for all growth processes is predetermined. Where the limits lie is uncertain. They are flexible. For example, they can be pushed back by technological innovations. On the other hand, the degradation of ever more ecosystems lowers this particular limit. What is certain is that the limits exist and have effects. At this point, the report introduces the term 'ultimate carrying capacity'. It comes originally from shipbuilding, where it denotes the highest allowable weight of freight which a ship can take on without sinking. From shipbuilding the term spread to scientific ecology. Here, it means the number of individuals from one species which an ecosystem can feed and therefore tolerate without negative impacts on its functioning capacity.

Given that the Earth is finite, two alternative courses of action are possible. One is to learn to live within the finite limits, to accommodate ourselves to them in order to remain within the carrying capacity, by halting growth, reducing the burdens on the system and achieving a state of equilibrium – before the limits are reached. The alternative is *not* to accept those limits, to ignore them in the hope of later being able to push them out further, and (if that doesn't happen in time) to 'overshoot' them. In which case, the end result would be an accommodation to those limits which would be enforced by nature rather than voluntary. Such an accommodation would be (literally) catastrophic: it would take the form of collapse.

The computer model does not describe this collapse. Nor is the 'period of overshoot' described in detail. In the sober language of the report the collapse is referred to as "a rather sudden and uncontrollable decline in both population and industrial capacity". Instead of painting a scenario of natural catastrophes, social disruption, hunger, chaos and genocide, the report identifies the factors most likely to provoke a collapse as a result of unbridled growth: depletion of the Earth's non-renewable resources, a leap in population growth and the resulting food shortage, and the Earth's declining capacity to absorb pollution. The overloading of the atmosphere with CO_2 is mentioned only in the margins. The authors warn that the burning of fossil fuels must be constrained before a serious "ecological or climatological effect" results. They discuss whether the then relatively new technology of atomic energy can contribute to that constraint. Probably not, is their answer, as it would produce "yet another kind of pollutant – radioactive wastes", which would be added to the "many disturbances man is inserting into the environment at an exponentially increasing rate".

Collapse is not – yet – inevitable. There is still a chance to change course. In defining what that new course should be, the report once again employs the term "sustainable": it talks of stabilising population, agricultural and industrial production and pollution within the "carrying capacity of the planet" in order to achieve a "sustainable state of global equilibrium". Such a dynamic balance could not be achieved by "technical solutions" alone. Rather, it would necessitate a combination of technical solutions and a fundamental change in the global human value system. In the past a whole culture had "evolved around the principle of fighting against limits rather than learning to live with them". The alternative to the "growth-and-collapse behavior mode" would be "to manage resources wisely". And every new idea would need to translate into a "visible improvement in the quality of life": in the future, the report suggests, growth would have to move over into the sphere of the immaterial.

Something important has happened here. In the Report to the Club of Rome of 1972 the classic principles and the old language of *sustaining use* and *sustained yield* have reappeared. But now they are no longer applied only to a narrow field like forestry. Another *world* is possible. "It is possible", the Report to the Club of Rome states, "to establish a condition of ecological and economic stability that is sustainable far into the future. The state of global equilibrium could be designed so that the basic material needs of each person on earth are satisfied and each person has an equal opportunity to realise his individual human potential." A new stage in the long history of conceptualising sustainability has begun. Now it's about Earth politics.

'The characteristics of a spaceship economy adapted to protect the environment'.[3] This is the title of a full-page article which appeared on 19 November 1972, half a year after the publication of *The Limits to Growth*, in the influential newspaper *Neue Zürcher Zeitung*. Its author was the Swiss construction engineer and entrepreneur Ernst Basler. He had brought the metaphor of the 'Spaceship Earth economy' back with him from the USA.

In the academic year 1969/70 Basler had been a guest professor at M.I.T., where one year later the Report to the Club of Rome was born. The title of his lecture there had seemed fairly innocuous: 'Engineering Strategy'. But his topic was decidedly up-to-the-minute. It concerned environmentally sound construction planning, or the adaptation of construction to the needs of a 'Spaceship Earth economy'. This metaphor had already been circulating at M.I.T. and other American think tanks for a couple of years. In his 1972 article Basler explained it. Like a spaceship, planet Earth is a closed and limited system. Its biosphere is finite and cannot be expanded. It follows that human economic activity must not exceed specific upper limits. For this reason a state of equilibrium without quantitative growth urgently needs to be achieved – no matter how impossible that might currently seem. Basler based his article on this model and on the just-published report to the Club of Rome. But in one sub-heading he broke new ground – *Erstes Merkmal: Nachhaltigkeit* ('The first characteristic: sustainability').

A complicated and exciting sub-plot reaches its conclusion here. Up until this point, the entire lexical field of sustainability had been established in German – *nachhaltig, Nachhaltigkeit*, etc. – but its application had been restricted to the narrow field of forestry. In English meanwhile the terms 'sustained yield' (also applying only to forestry) and now 'sustainable' (referring to the Spaceship Earth economy as opposed to the growth-and-collapse economy) had established themselves. Basler recognised the inner connection between the idea of the Spaceship Earth economy and *Nachhaltigkeit* in forestry.

He was the first person to bring together the futuristic spaceship metaphor, the highly topical debate about the limits to growth, and the old concept of *Nachhaltigkeit*, the basic principle of managing a forest. Ernst Basler had only recently, and as it were accidentally, learned of the forestry term himself in a professional exchange with Ulrich Zürcher, a high-ranking official of a Swiss industrial association. Zürcher was a graduate of the Forestry Faculty at the ETH Zurich; his doctoral thesis had been on sustainable forestry. The conversation had electrified Basler. The old concept for the closed economic systems of the cameralist central European territories was a perfect match with the brand-new idea of an economy for a finite planet. "The analogies", he wrote, "are so striking that the use of the term practically forces itself on you."

The newspaper article was an excerpt from the book *Strategy for Progress* which Ernst Basler had just published with a small Swiss publishing house. The book opens with a dedication to his three children. "In the hope that handing on to the next generation an unspoiled, stable and sustainable world may become a central concern of our culture." In the words of a father to his children, the spirit of the Brundtland formulation is anticipated here (and with more poetry, and more feeling).

". . . a durable and habitable planet"

The dominant image of the age we live in is that of the earth rising above the horizon of the moon – a beautiful, solitary, fragile sphere which provides the home and sustains the life of the entire human species. From this perspective it is impossible to see the boundaries of nations and all the other artificial barriers that divide men. What it brings home to us with dramatic force is the reality that our common dependence on the health of our only one earth and our common interest in caring for it transcend all our man-made divisions.[4]

These are the words with which the Canadian Maurice Strong, Secretary-General of the Conference, opened the United Nations Conference on the Human Environment in Stockholm on 5 June 1972. What the image did not reveal is that the human species has colonised even the furthest recesses of nature on the planet, and that all of these people have their needs, their demands, their picture of the good life. It was essential to bring both of these perspectives, from above and from below, into balance: Earth politics as global domestic policy. Strong expressed this dream, this version of globalisation as the idea of a durable and habitable planet.

The era of Earth politics began in 1968. Until then, the UN had shown little interest in the environment. Only when residues of the pesticide DDT were found in the body fat of Antarctic penguins, when increasing numbers of oil slicks polluted seas and beaches, and when Rachel Carson's book became a worldwide sensation did a consciousness of the global dimension of environmental problems seep into the multi-storey glass palace in New York.

Resolution 2398, passed at the beginning of December 1968 by the UN General Assembly in New York, noted that "the relationship between man and his environment is undergoing profound changes in the wake of modern scientific and technological development". These changes presented "unprecedented opportunities", but also "grave dangers", and in particular a "continuing and accelerating impairment of the quality of the human environment".

Human environment was the key term at this stage. It was understood to mean what surrounded human beings, their habitat: the life-supporting systems, that is, the biosphere and the ecosystems, but also nature as modified by Man. The

Stockholm Declaration would later open with the words "Man is both creature and moulder of his environment."

The initiative for Resolution 2398 came from Sweden. There, at the University of Uppsala, a soil scientist named Svante Odén had recently made a discovery which had set off a public outcry: the soils, forests and waterways of Sweden were being damaged by acid rain, caused by clouds of sulphur dioxide emitted originally in the industrial regions of Britain and Germany but travelling well outside those territories. The anger over the pollution of their fishing waters – angling is a national sport – and the concern for the survival of the native forests – still an important national asset – mobilised the whole country.

The General Assembly of the UN approved the Swedish resolution and decided to call an international conference on the human environment for 1972. It also accepted the Swedish proposal for the conference to be held in Stockholm.

During the preparations for the conference in Stockholm, climate change reached the UN agenda. U Thant, the Burmese Secretary-General of the UN, warned in 1970 that "a continued increase in excess of unabsorbed carbon dioxide could have a catastrophic warming effect, melting the polar ice, changing the marine environment and creating flooding on a global scale".

Frustrated at the slow progress of the preparations for the conference, in that same year U Thant appointed the Canadian Maurice Strong as a Special Envoy. Who was this man, who more than any other individual has supported, facilitated, shaped and driven forward the concept of 'sustainable development', from its beginnings, through the Rio conference, and up until today?

A photo of the closing ceremony of the Stockholm conference shows a middle-aged man in a business suit, with sideburns, hair receding at the temples and a small moustache. He is shaking someone's hand, bowing slightly, and with an engaging smile. At that time Strong was 42, and already seriously rich. He grew up the child of working-class parents in the Canadian province of Manitoba, and had experienced the misery of the Depression era at the sharp end, but had also been imbued with the ideals of the labour movement. His first step on the career ladder was a job in a trading post of the Hudson's Bay Company in the Canadian polar region. There he traded with the Inuit, learned their language and developed a fascination for their culture, not least for their ability to make the most of their sparse resources. His steep ascent of the ladder took place in the oil industry. At 35 he was already President of a leading Canadian energy company. Strong was at one and the same time both a shrewd businessman and a fervent idealist. As a young man already he had become an enthusiastic supporter of the ideal of closer understanding between nations and the Charter of the UN. Through his work for the Canadian YMCA he made contacts in the World Council of Churches. From 1966 he ran the development agency of the Canadian government, and learned about the problems of the developing world at first hand. Now he used his network of contacts and his skills as an entrepreneur in the preparation of the UN conference in Stockholm.

Only one Earth. That was the motto of the mammoth two-week conference. It was devised by the British ecologist Barbara Ward, no doubt influenced by the space photos, and perhaps inspired by the ancient *unus mundus* spirituality. 1,200 delegates from 113 countries gathered in the Folkets Hus, the modern glass and concrete headquarters of the Swedish trade union association. Additionally there came – for the first time in the history of the UN – countless representatives, experts and activists from NGOs, youth and grassroots organisations. But what was first revealed by the two midsummer weeks in the Swedish capital was the depth of the divisions between them.

On Sergels Torg in the city centre, angry groups protested against apartheid in South Africa and the French nuclear tests in the South Pacific. Thousands of young people marched through the streets behind a model of a whale to recordings of whale songs and cries of 'Val! Val!' (the old Norwegian whale hunters' call) to demand a ten-year moratorium on whale hunting. Street theatre performers picked up on this and started ironic drum-accompanied chants calling for a ten-year moratorium on the killing of people. In the tent city set up for the NGOs and youth groups on the abandoned airfield of Skarpnäck, members of the American hippie commune Hog Farm provided security and a whiff of Woodstock. (They had been responsible for 'soft security' at the legendary festival, too.)

The clash of principles and strategies continued in the conference hall. Delegations were constantly loudly walking out of the hall only to return, just as demonstratively, shortly later. The faces of the US delegates turned to stone when the Chinese called for the condemnation of the 'ecocide' – the defoliation of the forests by chemical weapons – in Vietnam, and all the more so when the host, the Swedish Prime Minister Olof Palme, joined them. A US scientist arguing for strict population control had the microphone pulled from his grasp by a Brazilian economist who angrily accused him of promoting 'genocide of the unborn'. A delegate from the Ivory Coast said mockingly that his country would happily put up with US levels of environmental pollution if they could have US levels of prosperity. No growth? Environmental protection is a luxury we cannot afford! Only one Earth? We insist on our sovereign right to exploit our own domestic resources!

A dark-skinned woman in a colourful sari then stepped on to the podium and raised the debate to a new level. With the authority that came from having been elected by a wide majority in the most populous democracy in the world, and from being able to claim some success in the fight against hunger in one of the poorest countries on Earth, the Indian Prime Minister Indira Gandhi was the only foreign head of state to attend the conference. *Garibi Hatao* – "abolish poverty" – was her slogan back home (though some thought it mere populism). And it provided the leitmotif for her forceful speech in Stockholm.

> On the one hand the rich look askance at our continuing poverty – and on the other they warn us against their own methods. We do not wish to impoverish the environment any further and yet we cannot for a moment forget the grim poverty of large numbers of people. Are not poverty and need the greatest polluters?

Indira Gandhi talked of the inhabitants of Indian villages who were forced every day to comb through nearby jungle in the search for food and the means of subsistence. And she asked, "Must there be a conflict between technology and a truly better world? Or between enlightenment of the spirit and a higher standard of living?" To which she gave her own answer: "The inherent conflict is not between conservation and development but between environment and the reckless exploitation of man and earth in the name of efficiency." This was directed as much at the capitalist West as at the socialist East. "All the -isms of the modern age, even those which in theory disown the private profit principle, assume that man's cardinal interest is acquisition. The profit motive, individual or collective, seems to overshadow all else. This overriding concern with Self and Today is the basic cause of the ecological crisis. . . . Modern man must re-establish an unbroken link with nature and with life. He must again learn to invoke the energy of growing things to recognise, as did the ancients in India, centuries ago, that one can take from the earth and the atmosphere only so much as one puts back into them." The dazzling climax of her speech was a quotation from a 3,000-year-old Indian text: "In their Hymns to the Earth, the sages of the Atharva Veda chanted: 'What of thee I dig out, let that quickly grow over. Let me not hit thy vitals or thy heart.' So can man himself be vital and of good heart and conscious of his responsibility."

Her speech was anything but a plea for development regardless of environmental consequences. Nevertheless, in the decades which followed, Indira Gandhi's message was all too often reduced to the simplistic assertion that 'poverty is the worst form of environmental pollution'. Taken out of context, this slogan was used for decades to justify the blocking of efforts to introduce international protection of the environment and of resources – in the name of 'development'.

But what was meant by this term at that time? In the 1950s, in spite of all the conflicts at the UN, a broad consensus had crystallised among the governments of the industrialised countries and among the elites of the former colonial states.

164

Development meant becoming like the developed countries. Achieving as swiftly as possible the GDP and per capita income of the industrialised countries and the living standards of the American middle class by means of an industrial infrastructure created – if necessary – overnight and out of nothing. But now the secondary meaning of the word 'development' in American usage began to make itself felt. For there it always means in addition the 'opening up' of a property or a region for commercial exploitation. Neither consideration of the environment nor concern for the improvement of the circumstances of the poorest of the poor had a place in this strategy. Rising wealth would 'trickle down' and would relieve poverty without further intervention, and at the same time it would make resources available for protecting the environment. This model of development dominated thinking in this area from the 1960s onwards, and still drives the disastrous course of 'catch-up development' today.

Notwithstanding the misuse of some of her arguments, in Stockholm Indira Gandhi initiated a global debate about the relationship between poverty and ecology. The idea of environmental protection combined with the imperative of combating poverty, and both were transformed into the idea of sustainable development. The closing Declaration already put down important markers: "Man is both creature and moulder of his environment, which gives him physical sustenance and affords him the opportunity for intellectual, moral, social and spiritual growth." It links together "the fundamental right to . . . a life of dignity and well-being" with the "solemn responsibility to protect the environment for present and future generations" (Principle 1). "The capacity of the earth to produce vital renewable resources must be maintained and . . . restored or improved" (Principle 3). Essential for this purpose is "economic and social development . . . for ensuring a favourable living and working environment for man. . . and for the improvement of the quality of life" (Principle 8). All states "have the sovereign right to exploit their own resources and the responsibility to ensure that these activities . . . do not cause damage to the environment" (Principle 21).

Think globally, act locally – this slogan was first heard in Stockholm. It was coined by the French-American microbiologist, doctor and environmental researcher René Dubos, then 71 years old. He explained it in a speech during a plenary session. "As we enter the global phase of social evolution it becomes obvious that each one of us has two countries – his own and the planet Earth." For all the problems besetting Spaceship Earth a global approach is necessary. But local solutions, directed of course by local interests, are essential for the specific problems of every human settlement. The two approaches are complementary. A rational loyalty towards the planet as a whole need not interfere with the emotional attachment to each other in a community, nor to a deep respect for diversity. We cannot feel at home on Earth if we do not continue to love and cultivate our own garden. And conversely we can hardly feel comfortable in our garden if we do not care for the planet Earth as our collective home.

Many of the speeches in Stockholm were filled with the sense that right there and then something momentous was happening, a turning point of world-historical significance, the dawn of a new age. Maurice Strong called it the "age of environment". In their joint declaration, the NGOs spoke of a "revolution in thought", comparable to the Copernican revolution. Then, "men were compelled to revise their whole sense of the earth's place in the cosmos . . . today we are challenged to recognise as great a change in our concept of man's place in the biosphere . . . Our survival in a world that continues to be worth inhabiting depends upon translating this new perception into relevant principles and concrete action." Barbara Ward, the British ecologist and philosopher, picked up this thread and span it on further. In the oldest elaborate human civilisations, which emerged from the neolithic revolution, the militarism of the state, the luxury and rapacity of the leaders had gained the upper hand. But whether in India, China or the Middle East, the great ethical systems of mankind – from the benign wisdom of Confucius to the passionate social protest of the Hebrew prophets – had expressed a deeper moral reality: that we live by moderation, by compassion, by justice, that we die by aggression, by pride, by rapacity and greed.

Maurice Strong had said in his opening speech that "our purpose here is to reconcile man's legitimate, immediate ambitions with the rights of others, with respect for all life-supporting systems, and with the rights of generations yet unborn." The search for a term which would express the synthesis of environment and development, of the present and the future, had begun.

"Behold, I make all things new"[5]

In the summer of 1968, the World Council of Churches, an association of Christian churches and communities representing about 500 million members worldwide, held its seven-yearly Assembly in Sweden. The theme was 'Behold, I make all things new'. The days of debate on the campus of the University of Uppsala – Linnaeus's alma mater – were stormy. To the dismay of the orthodox and evangelical delegates, the political and social grass roots movements of the time spoke out in messianically radical terms. 'Combat racism!' and 'bread for the world!' and the theology of liberation all took centre stage. The US folk bard Pete Seeger sang *We shall overcome*, the hymn of the American civil rights movement.

One of the key speakers was Margaret Mead from the USA, representing her Episcopal Church. The 67-year-old ethnologist and anthropologist was an icon of the 1968 generation, and not just in the USA. She drew on the findings of the ethnographic research she had carried out as a young woman on the South Pacific island of Samoa to call for radical change in western societies: they should recognise the diversity of cultures in the world as an asset; should focus on the quality of life rather than on living standards; and should allow a more open sexual morality. 'Learning to live in one world' was her motto already in 1945. Now she vehemently criticised the churches' distance from social reality. She demanded Christian responses to the scientific and technological revolutions and the global crisis they had ushered in. The Council reacted. It set up a commission of enquiry. Its topic was 'The Future of Man and Society in a World of Science-based Technology'.

Over the course of the next few years, the Australian biologist Charles Birch became the intellectual leader of this commission. Birch was a scientist of global standing, a practising Christian, a figurehead of the Australian protest movement against the Vietnam war, and a member of the Club of Rome. It was through this that he made contact with the young Norwegian Jorgen Randers, at that time working on the report on the limits to growth at MIT in Cambridge.

In the summer of 1974 the commission met in Bucharest. For a couple of days the Balkan metropolis, a testament in depressing grey concrete to the megalomaniacal dictator Ceauşescu, was the stage for a colourful workshop on the future. In the featureless and dreary hotel which was the conference venue Margaret Mead energetically developed her ideas on the primacy of the quality of life over living standards as measured in material goods. One working group concerned itself with escape routes from the vicious circle of global armaments and defence research. Another working group, led by Charles Birch, was discussing *The Limits to Growth* with Jorgen Randers. Just like at the UN conference two years earlier in Stockholm, the term 'limits' was polarising the participants. In this Christian body, too, the representatives of the developing countries insisted emphatically on the primacy of the issue of justice. Indeed they were hostile to the idea of 'limits'. They argued that the rich countries had had their share of growth. Now it was their turn. Don't talk to us about limits to growth, they said. What we need is to grow just as the rich countries have grown.

At a coffee break Birch discussed the situation with Randers. Both had the feeling that the workshop was getting nowhere. At that point, Birch recalled later, Randers said to him: "We have to find some phrase other than 'limits to growth' that is positive in its impact. Limits has a negative connotation. Other suggestions like 'stationary state', an 'equilibrium society' and a 'steady-state society' are too static." Then he asked, "What about 'the ecologically sustainable society'?" They took this suggestion back into the discussion session with them and found that it met immediately with a warmer reception.

In the closing document of the Bucharest conference the new term appears in a central position. The starting point is a differentiated consideration of 'quality of life'. The purpose of all thinking about the future of humanity is to achieve a higher quality of life. However, the "swelling material activity on our fragile, finite planet" has reached the point where more material production in "the rich segments of the world" will no longer lead to the satisfaction of the non-material dimensions of the quality of life. The rest of humanity, though, the poorer part, is still dependent on material expansion. "Thus today the world-wide quality of life will be increased by material growth among the poor and by stabilisation and possibly contraction among the rich." The material conditions of life for the poor have to be improved. But the increase in productivity which that requires must not lead to the collapse of, for example, agriculture. "The goal must be a robust sustainable society where each individual can feel secure that his/her quality of life will be maintained or improved." The report then describes the desired state or condition of sustainability under four aspects. It entails, first, "an equitable distribution of what is in scarce supply and a common opportunity to participate in social decisions". A society will not be sustainable unless, second, "the need for food at any time remains well below the global capacity to supply it, and unless emissions of pollutants are well below the capacity of the ecosystem to absorb them". Third, a social organisation will be sustainable only as long as "the rate of use of non-renewable resources does not outrun the increase in resources made available through technological innovation". And finally, a sustainable society will require "a level of human activity which is not adversely influenced by the unending, substantial and frequent changes in the global climate".

The declaration analyses the timescale for the necessary changeover to a sustainable society with cool precision. "To avoid catastrophe in an expanding system, efforts to control the growing impact must be begun before the anticipated levels approach the unacceptable. Normally this implies action at a time when the present situation is still acceptable, often desirable. . . . Even at the risk of being unable to complete the transition into a sustainable society, we must have the faith to take the first step, and to believe that human ingenuity can overcome the obstacles." 'We shall overcome' – the great dream of the 1960s came to life again during the days of the Bucharest conference.

The *sustainable society* concept caught on at the World Council of Churches. In a plenary session of the next Assembly, which was held in Nairobi in 1975, Charles Birch gave a keynote speech on the findings from Bucharest. The organisation took the decision to revise and expand its vision statement. A "just and participatory" society became a "just, participatory and sustainable" society. Conceptualising sustainability had taken a big step forward. What had seemed a marginal attribute when it appeared in the Report to the Club of Rome had now been transformed into a clearly defined political term. *Sustainable* had now been recharged with a fresh set of ideas relating to justice, participation and quality of life.

'Only one world', and the image of the blue planet, can be interpreted in a radically different way. The Earth is a homogenous space, desirable, permeable and free from barriers, potentially subject to control down into its furthest recesses, open for boundless economic expansion and the globalisation of the market. In the 1990s a credit card firm targeted new customers with the image of the Earth from space and the slogan 'MasterCard – the world in your hands'.

In the years of Earth politics a radical counter-utopia took shape. At first confined to small academic circles, it quickly moved – in leaps and bounds – onto the global stage. In the lecture theatres and seminar rooms of the University of Chicago, economists such as the Austrian Friedrich von Hayek and the American Milton Friedman adjusted the classic liberal doctrine of the free market to take account of changed conditions. This neoliberal doctrine formed a triangle constructed around the three points of privatisation, deregulation and reductions of social security. The neo-liberal thought leaders and their followers now started to put the theory into practice. Their instrument was what the Canadian writer Naomi Klein in 2007 called *The Shock Doctrine*.[6] Following the putsch there in 1973, Chile became the first testing ground. In the days and weeks which followed the bloody overthrow of the socialist Allende government, the economy of the country (which was paralysed by shock) was turned upside down. Under the protection of the terrorist Pinochet regime, all areas of life were subjected to the rules of the free market.

Two years earlier, in 1971, what was later to be called the World Economic Forum was founded in Europe. The original aims appeared to be modest. At annual meetings in Switzerland, European firms would be able to acquaint themselves with the latest management practices from the USA. The 'mission' was soon expanded to 'undertaking global initiatives to improve the state of the world'. Access was restricted to firms with a turnover – usually – of at least five billion dollars, and then gradually widened to include politicians, the media, and celebrities. The wintry gatherings in the glitzy world of the Swiss Alps provided an umbrella of glamour and credibility for the economic strategy of neo-liberalism. 'Davos' and 'Chicago' became points around which crystallised the globalisation of a radically market-driven form of capitalism.

The neoliberal doctrine clashed at all points with the principles and the philosophy of sustainability. *Only one Earth* is a call for the Earth and its resources to be understood as the common property of humanity rather than as the private property of a few. *Sustainability* requires a web of ecological, economic and social regulations. *Sustainable development* is a strategy for the extension of participation and of a high quality of life, of access to a fuller life, to everyone. The collision between these two philosophies and political strategies still continues today. It is the real clash of cultures going on in the world of the 21st century.

Gaia

During this phase of Earth politics, in the middle of the 1970s, a new scientific worldview caused a considerable stir. The Gaia hypothesis offered a breathtaking new perspective to ecological research. At the time it was very controversial. Today, as Earth-system science or geophysiology, it is an established discipline in the mainstream of the sciences.

> When the Earth was first seen from outside and compared as a whole planet with its lifeless partners, Mars and Venus, it was impossible to ignore the sense that the Earth was a strange and beautiful anomaly.[7]

Once again, the fascination of the planetary perspective. For the British researcher James Lovelock, this had been the trigger. Since 1965 he had been working for NASA at the Jet Propulsion Institute in Pasadena, California, researching into the solar system. The beginnings of space travel had re-ignited interest in one of the greatest quests of human history: the search for extraterrestrial life. It was given additional impetus by a vision which – although unspoken – was part of what drove the entire NASA project: the colonisation of other planets and the idea of 'Terraforming', the founding of a second Earth. The primary object of scientific curiosity in this area was the red planet. Lovelock's thesis was that the chemical composition of the atmosphere on Mars would enable us to draw conclusions about the existence of life there. The metabolic processes which accompany life, in whatever form, would be bound to leave traces in the atmosphere. He managed to gather sufficient data on the geo-chemistry of Mars's atmosphere, and came to the conclusion that no life existed in the atmosphere of our neighbouring planet, consisting as it does almost entirely of CO_2. His findings have still not been overturned.

But this disappointing result (for his employers) did not satisfy Lovelock. In the course of his research he had asked himself what were the natural conditions which were absolutely essential for life, and he had shifted the direction of his enquiry back towards Earth. When he compared the measurements from Mars with the data for Earth he was struck by astounding differences. For the Earth's atmosphere (especially in respect of the high proportion of oxygen) was a crazy cocktail of gases, one which – though exceptionally conducive to life – could not be explained on the basis of chemical and physical processes alone. And what was even more astounding was that this highly reactive mixture had apparently

170

remained stable over millions of years. The gases were therefore clearly being regulated and replenished, and the pump responsible for this had to be the living organisms themselves. It followed that the Earth has a perfect system of self-regulation. It behaves like a living system. In 1968 Lovelock published his theory for the first time in a scientific journal.

At around the same time, on the American east coast, the evolutionary biologist Lynn Margulis was investigating not the distant realms of the solar system but the world of microorganisms. She was researching into why bacteria and other single-celled organisms were the only forms of life which had existed on Earth for a billion years. How had these life-forms organised themselves into functional and efficient communities? How did life evolve further from this microbial biomass? In 1965 Lynn Margulis first made public her theory that the cells of all animal and plant organisms developed out of the fusion of different species of bacteria. Symbiosis – that is, mutual dependence, cooperation and union between different organisms; or, in other words, protracted intimate relations between strangers – was in this view a fundamental driver of evolution. The diversity of life developed from symbiotic communities of different organisms, from their co-habitation, fusion and amalgamation. This theory confirmed the Darwinistic model according to which all organisms had common ancestors. It refined the view, already expounded by Darwin, that natural selection favoured not necessarily the strongest individuals but rather the best-adapted and most cooperative. At the same time it attacked the widespread dogma among Darwinists and Neo-Darwinists that competition and the life-or-death struggle within and between the species was the overriding mechanism of natural selection, and thus of survival.

From 1971 onwards Lovelock and Margulis worked together. Out of their joint brainstorming emerged the hypothesis to which Lovelock gave the name of the Greek goddess of the Earth, *Gaia*. In 1979, Lovelock defined Gaia as "a complex entity involving the Earth's biosphere, atmosphere, oceans and soil; the totality constituting a feedback or cybernetic system which seeks an optimal physical and chemical environment for all life on this planet". What had led the two researchers to this hypothesis?

One starting point was the global climate. Although solar radiation on to the Earth has increased over the lifetime of the planet by about 30 percent, the temperature has remained roughly constant. What causes this thermostatic effect? To take one example: when the Sun shines more intensely on the surfaces of the seas and oceans, the plankton population increases faster. This leads to increased emission of DMS (Dimethyl sulphide) into the air. Particles of this substance act as condensation nuclei, forming the water droplets that make up clouds. But thicker clouds contribute to the cooling of the lower layers of the atmosphere. It gets cooler again. Another observation concerned the relatively constant level of salt in the oceans. The ceaseless flow of dissolved salts delivered by the rivers from the soils and rocks of the continents over the course of millions of years ought to have led, slowly but surely, to higher salt levels in the seas – to a point where all life would have died out. But this rise in salinity has not occurred. This is because diatoms, a type of algae, store salt in their shells and deposit it as sediment on the sea bed. They keep the marine habitat habitable. The issue of the peculiar chemical make-up of the atmosphere was central. This could only be explained by the fact that all living things continually release reactive gases into the atmosphere, the soil and the water, and in turn take back others. The totality of these metabolic processes accounts for the high proportion of oxygen and thus for the life-supporting composition of the atmosphere. All of these phenomena have been maintained over endless stretches of time. They take place, and have always taken place, with a remarkable precision.

The deduction which follows from this in the Gaia hypothesis is that the Earth is not a dead ball of rock on which life has to adapt reactively to the geochemical processes of an inert environment. Life on Earth in its totality is instead a proactive factor: lifeless matter is shaped, influenced, even directed by it. Life itself provides for the life-supporting or life-conducive conditions on the surface of the Earth. All of its metabolic processes, micro-organic, vegetable and animal, work together with inanimate nature. It regulates and modulates it constantly and thereby maintains intact the conditions for its own existence. Symbiotic communities – algae and clouds, fungi and rocks – together weave this web of life and maintain it. In this holistic worldview, the planet is seen as a highly complex living body fed by solar energy, an interactive, self-regulating, networked system with a colossal number of feedback effects. The Earth, according to the Gaia hypothesis, behaves like a gigantic living organism.

This view puts the history of evolution and the laws of natural selection in a fundamentally different light. "There was no purpose in this," Lovelock wrote, "but those organisms which made their environment more comfortable for life left a better world for their progeny, and those which worsened their environment spoiled the survival chances of theirs. Natural selection then tended to favour the improvers."

So what role does humankind play in this complex game? The Gaia hypothesis argues that humankind is an inseparable element of the web of life. With every breath, with every mouthful of water, every one of our cells is in contact with the Earth and with the cosmos. Our intellect has the responsibility of contributing to the improvement of the conditions of life, perhaps even one day of protecting the Earth against dangers from space. But our fantasies of omnipotence will not be tolerated. As Lovelock says, "Gaia, as I see her, is no doting mother tolerant of misdemeanours, nor is she some fragile and delicate damsel in danger from brutal mankind. She is stern and tough."[8] Instead of drawing attention, like the astronauts, to the ethereality and fragility of the planet, the Gaia hypothesis emphasises the sturdiness and resilience of nature, and thus its capacity for resistance. To the destructive activities of humankind – Lovelock speaks of the three Cs: cars, cattle, chainsaws – Gaia responds resolutely and rigorously, with merciless feedbacks which could lead – in the interests of life overall – to the decimation, even to the elimination of the human species. If a collapse is to be averted, the rethinking has to go very deep.

The term *biodiversity* came into circulation in 1985. The American biologist who coined it, Edward O. Wilson, dramatically raised awareness of the eminent importance of biodiversity for the stability of ecosystems. Which contemporary threat, he asked, would prove the worst legacy for our descendants in 1,000 years' time? His answer was that it would be neither the depletion of the energy reserves, nor economic collapse; nor even a limited nuclear war or a totalitarian takeover of government. As dreadful as such catastrophes would be for us, their consequences could be remedied within a few generations. "The one process ongoing in the 1980s," Wilson wrote, "that will take millions of years to correct is the loss of genetic and species diversity by the destruction of natural habitats."

In the 1990s the metaphor of the *ecological footprint* entered into the discourse of sustainability. The ecological footprint is "the amount of biologically productive land and sea area necessary to supply the resources a human population consumes, and to assimilate associated waste". It provides an objective measure for the increasing consumption of natural resources and the 'overshooting' of the limits of ecological carrying capacity – in short, for the plundering of the planet.

Unexpectedly, and hearteningly, *Gaia* put a new face on the myths of *mater terra*. A whiff of *natura naturans* wafted up to us from Spinoza's time. A new spirituality, in tune with nature, opened up the possibility of the re-enchantment of the world. At the same time, the new worldview served to further the solution to the most urgent problems facing us. The thinking about ecology and sustainability was developing in both depth and breadth.

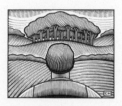

CHAPTER THIRTEEN

Earth politics II: moving on to the next level

Living Resources

> The aim of the World Conservation Strategy is to help advance the achievement of sustainable development through the conservation of living resources.[1]

'Sustainable development' is an expression coined by conservationists. It was launched into the wider world on 5 March 1980. On that day, a text stretching to barely five pages entitled 'World Conservation Strategy' was presented simultaneously in 35 capital cities. The subtitle was 'Living Resource Conservation for Sustainable Development'. This was the first time that 'sustainable development' had been used as a consciously chosen and clearly defined composite term. The introduction stated that "humanity's relationship with the biosphere . . . will continue to deteriorate until . . . *sustainable* modes of *development* become a rule rather than an exception."

Almost 1,000 experts had worked on this 'world conservation strategy': conservationists, ecologists, foresters and development specialists. The coordinators were the International Union for Conservation of Nature (IUCN) and the United Nations Environment Programme (UNEP). Its beginnings went back to the year of the Moon landing.

In 1969 the General Assembly of the IUCN met in New Delhi. The closing statement of the conference begins by considering the cosmos, nature and humankind, and declares that "the splendour of this earth derives from its sunlight, its beautiful green cover, its interdependent fauna and flora, and from the diversity of its landscapes"; and that "since the beginning of its existence, the people of this earth even when poor in material possessions have found life richly worth living because of these natural assets"; and finally that "man, himself a product of the evolutionary system, is dependent on the stability and self-renewing properties of his environment."

Like the UN and the World Council of Churches, the IUCN was a child of the post-World War II period. It too was suffused with the desire, following an era of global warfare and genocide, to ensure stable, humane, and just relations in order to make life worth living in an integrated world. Representatives of the governments of over 100 countries and countless NGOs were joined together in this global community of conservationists. The World Wildlife Fund (now known outside North America as the World Wide Fund for Nature) was set up with the help of IUCN as a fundraising organisation. Among its supporters were European royalty like Britain's Prince Philip and Prince Bernhard of the Netherlands. At lavish charity galas in New York's Waldorf Astoria or in London's Ritz, they made the WWF and its logo, the panda, famous around the world.

On the ground, the work of the IUCN addressed classic conservation issues: red lists of those species threatened with extinction, moratoria on the hunting of wild animals, the establishment of extensive protected zones in the few remaining areas of wilderness. 'Serengeti shall not die' was a typical campaign from that time. It was hoped that the creation of nature reserves in tropical countries would prevent the extermination of big game by trophy hunters, ivory traders and poachers. The new elites in the former colonies mocked these 'big game hunters with a conscience' as colonialists by other means.

Around the middle of the 1960s, the top management of the IUCN and WWF recognised the need for new strategic thinking in conservation. They had to move on from a firefighting strategy involving the dramatic rescue at the last minute of threatened natural icons, and from a conservation strategy based around animal reservations in which the human inhabitants were perceived only as an irritation. Constructive cooperation and action rather than restrictions and prohibitions. Strategic long-term programmes rather than ad hoc reaction. Focusing on solutions rather than on problems.

The General Assembly in New Delhi in 1969 moved the concept of *quality of life* into the centre of the IUCN programme. By this was meant the "physical, educational, social and aesthetic values which add richness, meaning, and satisfaction to human experience". In this context the opportunity was taken to define more precisely what was meant by *conservation*, the guiding concept of the organisation. A purely 'conserving' conservation, one which absolutely prohibits any utilisation of nature for human benefit, was no longer in keeping with the times. It needed to be supplemented by *management*, and that would include "the survey, research, administration, preservation, utilisation . . . of air, water, soils, minerals and living species". Conservation and the highest sustainable quality of life belonged together.

The IUCN worked on its new programme for ten years, from 1969 to 1979. The list of those involved reads like a *Who's Who* of the international environmental movement. Gerardo Budowski, a Venezuelan of German descent and expert on tropical forestry, was the Director General of the organisation. Naturalists like Raymond Dasmann of the University of California, Wolfgang Burhenne, an environmental lawyer working for the German government, Lynn Margulis, the Gaia researcher from Boston, and Albert Baez, chair of the IUCN education commission and father of the folk singer Joan Baez, were among the contributors. Most of them were familiar with the basic principles of the science of ecology. Many of them used the *sustained yield* concept from sustainable forestry. In the course of their work on the programme the ecologists developed a better understanding of the need to reduce poverty.

Their starting-point was a thorough analysis of renewable resources. In contrast to non-renewable resources (i.e. fossil and mineral raw materials), renewables are always living resources, vegetable or animal organisms, living creatures. Not dead matter, like fossil resources. They live. They are subjects, not objects. They create processes. Even the air that we breathe and the soils on which we cultivate our food are the products of living organisms. They are renewable if conserved; and they are destructible if not. So they must never be allowed to be depleted so far that they lose the capacity to reproduce. Their habitats, the ecosystems to which they belong, must remain permanently intact. If not, the species will die out. And they will then no longer be regrowing, but irretrievably lost. Lost, too, will be the benefits that human beings derive from their existence. This fact seems simple and straightforward. But it has far-reaching implications. It overturns the logic of the fossil-fuel age. When the resources have been used up, you can't simply tap into new deposits; you have to wait until they have regrown. The conservation of living resources requires diversity. The text of the World Conservation Strategy still speaks of genetic diversity. Only a few years later, the term 'biodiversity' begins to take over. The basic realisation was that diversity of species, and genetic diversity within species, are prerequisites for the stability and resilience, the capacity for resistance, of ecosystems. Monocultures are never sustainable. Maintaining diversity is keeping options open for the future.

A new approach to living resources required a new language. The key terms of the 'World Conservation Strategy' are *sustain* and *sustainable*. Earth is the only place in the universe known to sustain life. The first sentence of the Foreword becomes the starting-point for an analysis of the global crisis. "[H]uman activities are progressively reducing the planet's life-supporting capacity at a time when rising human numbers and consumption are making increasingly heavy demands on it ... the combined destructive impacts of a poor majority struggling to stay alive and an affluent minority consuming most of the world's resources are undermining the very means by which all people can survive and flourish. ... Humanity's relationship with the biosphere (the thin covering of the planet that contains and sustains life) will continue to deteriorate until a new international economic order is achieved, a new environmental ethic adopted, the human population stabilises and sustainable modes of development become the rule rather than the exception."

The analysis of the interrelationship between development and conservation begins by defining its terms. What is development? It is "the modification of the biosphere and the application of human, financial, living and non-living resources to satisfy human needs and improve the quality of human life. ... For development to be sustainable it must take account of social and ecological factors, as well as economic ones: of the living and non-living resource base; and of the long-term as well as the short-term advantages and disadvantages of alternative actions." What is conservation? The 'World Conservation Strategy' defines it as "the management of human use of the biosphere so that it may yield the greatest sustainable benefit to present generations while maintaining its potential to meet the needs and aspirations of future generations". This definition of *conservation* anticipates right down to the choice of words the Brundtland formulation for sustainability of 1987.

Here then is the conceptual link which binds the *providentia* doctrine of medieval theology, the sustainability philosophy of the early Enlightenment (*conservatio*), the forestry term as developed at Tharandt (*sustained yield*), and the vocabulary of the Brundtland Report. Last but not least, the new definition incorporates the wisdom of ecology from the times of Linnaeus and Goethe. We are part of nature. We cannot and do not manage it. What we *can* manage is our own conduct towards the biosphere.

Rethinking development

Another pioneering report appeared in February 1980, at the same time as the *World Conservation Strategy*. It bore the title *North-South: A Programme for Survival*. The contributors, under the leadership of Willy Brandt, were for the main part politicians involved in peace and disarmament policy and experts in development policy and the fight against poverty. From their own particular perspective they also looked at the ruination of the biosphere. The Brandt Report contained a new definition of *development*.

The origins of the report lay back in 1977. In that year, Robert McNamara, the President of the World Bank, had asked the former German Chancellor Willy Brandt, together with an international commission, to analyse the fundamental problems of global development and to develop proposals to deal with them. At first Brandt had hesitated. He had not specialised in the politics of development, even if he had followed the process of decolonisation with interest and sympathy all his political life. Nor had he specialised in environment policy, even though he had attracted attention by making an issue of pollution and by calling for 'blue skies over the Ruhr region' in his first major campaign, the federal government elections in 1961. His real field of interest was the politics of détente between East and West. It was for this that he had received the Nobel Peace Prize in 1971.

What did characterise Brandt was that his political compass was set on solidarity and inclusiveness, both within his own society and within the international community. He saw the politics of social justice as a condition of peace. In his Nobel Prize lecture in 1971 he argued that "a durable and equitable peace system requires equal development opportunities for all nations". Two years later, to mark the admission of the Federal Republic of Germany to the UN, he gave a keynote address to the General Assembly. It contained his vision of what he would later term "global domestic policy"; it had a strong environmental component and was based on the Spaceship Earth metaphor. "It is probably not a coincidence that now that man has seen his planet from the depths of space he has become aware of the material and biological dependency of the inhabitants of this very small spaceship Earth ..."

In December 1977 the North-South Commission embarked on its work, under the leadership of Willy Brandt. Representatives of the South were in the majority: five from Asia, three South Americans and one African, working together with three Europeans and three North Americans. In a two-year marathon of conferences, consultations and meetings around the globe the Brandt Commission worked on its report. Its title, *North-South: A Program for Survival*, signalled a departure from the traditional division of the world into 'developed' and 'underdeveloped' countries. Instead, it took up the theme which had dominated the global debate on environment and development since Indira Gandhi's speech in Stockholm and placed it at the centre of its analysis: how can the deep divide between the rich countries of the North and the poor countries of the South be overcome? How can the Earth come together to form the *one* world so often invoked?

Brandt found a simple starting-point: the fact *that mankind wants to survive*.[2] In its risk analysis, his report follows the line taken eight years earlier by the 'Club of Rome'. It lists the burgeoning world population, the endangering of the environment through deforestation, the plundering of fish stocks, the pollution of air and water, the irreparable damage to ecosystems and not least climatic change with potentially catastrophic consequences caused by CO_2 emissions. However, the report points out more clearly than any of its predecessors the political dangers: the growing international tensions caused by competition for energy, food and raw materials, and (for the first time) the misuse of power by the elites and the fanaticism of the masses in the countries of the South. And this report, published almost 30 years ago, also contains an early warning of the possible collapse of the credit system, caused by the inability of debtors to pay and the consequent bank failures. Another burning current question anticipated in the Brandt Report of 1980 is 'peak oil'. "The oil stock of our planet has been built up in a long process over millions of years, and is being blown 'up the chimney' within only a few generations. Are we to leave our successors a scorched planet of advancing deserts, impoverished landscapes and ailing environments?"[3]

Allowing things to take their course is out of the question. The frightening prospects rule out a strategy of business as usual, or 'laissez-faire'. Nevertheless the report holds fast to the idea of "a hopeful future" and the vision of a better world. Against the spectre of wars fought for dwindling resources it holds up the simple truth that war cannot increase resources: "[T]here certainly is no military solution to the problems of energy and commodities."

The recognition of interdependence, of mutual dependence on the planet, must prevail. "We see a world in which resources are squandered without consideration of their renewal."[4] To counter this, the report calls for "alternative" solutions and "an orderly transition from a world economy and industry based on oil, to one that can be sustained through renewable sources of energy", and for the abandonment of a lifestyle based on excessive energy use.

The word *sustainable* is used at several points in the text of the Brandt Report. It speaks of *sustainable biological environment* and of *sustainable prosperity*. But above all the report introduces a radically new definition of 'development', and for this it goes back to the original meanings of the word.

Brandt writes in his foreword that "One must avoid the persistent confusion of growth with development." A development strategy which aims simply at expanding production, at growth in GDP or the standard of living (however measured) can no longer be the guiding principle. Rather, the idea that the whole world should be modelled on the highly developed countries should be dropped. The first and overriding aim must be that of an equitable distribution of income. Brandt opposes cultural imperialism and makes a plea for cultural and political diversity. "There is no uniform approach. There are different and appropriate answers depending on history and cultural heritage, religious traditions and economic resources, climatic and geographic conditions, and political patterns of nations."[5] When the focus is on the quality of growth, then "the creation of jobs and the basic needs of the poorest sections of society" are taken into account. "A people aware of their cultural identity can adopt and adapt elements true to their value-system and can thus support an appropriate economic development . . . We strongly emphasise that the prime objective of development is to lead to self-fulfilment and creative partnership in the use of a nation's productive forces and its full human potential." Development, Brandt says, means the "unfolding of productive possibilities and of human potential". Here *development* is finally uncoupled from the imagery of brute, capital-intensive industrialisation, of the production and consumption of goods, of purely economic growth. The concept is aligned once more – as it was for the philosophers of the early Enlightenment – with *unfolding*, and with personal growth.

A stunning definition of development was introduced in 1980 by the International Foundation for Development Alternatives, a network of activists mainly from the South, cooperating with the Brandt Commission:

> Development is the unfolding of people's individual and social imagination in defining goals, inventing means and ways to approach them, learning to identify and satisfy socially legitimate needs. Development, thus defined as liberation of human beings and societies, happens, or better, is lived by people where they are, that is, in the first instance, in the local space. . . There is development when people and their communities. . . act as subjects and are not acted upon as objects; assert their autonomy, self-reliance and self-confidence; when they set out and carry out projects. To develop is to be, or become. Not to have.[6]

Willy Brandt, too, communicated a vision of the future which he expressed as follows in the second report from his Commission three years later: "A new century nears, and with it the prospects of a new civilisation." At the minimum this will be a civilisation in which no child will have to starve any longer, and "no-one will have to see uncomprehending panic in . . . the clear, radiant eyes of children." In the reports of the North-South Commission the idea of sustainability begins to acquire a clear outline. It is more than an environmental policy concept, more than a development policy strategy, more than a plea for technological innovation. It is conceived of as a new model of civilisation.

In his memoirs, Brandt recalled a conversation he had with the President of Peru in the mid-1980s. The country was at that time undergoing a neoliberal 'shock therapy'. "We are being operated on without anaesthetic," the President had said to him. "They want us to feel the pain." In 1979 Margaret Thatcher had taken over government in Great Britain, in the following year Ronald Reagan had followed her in the USA. Both were passionate believers in the neoliberal economic doctrines of Milton Friedman. They made privatisation, deregulation and the fetishisation of economic growth into the guiding principles of politics around the globe. The race for dwindling economic resources accelerated. Sustainable Earth politics collided with the globalisation of radical free-market thinking. The struggle between these two principles is by no means over.

The Brundtland Formulation

In the autumn of 1983 the leadership of the UN appointed the Norwegian politician Gro Harlem Brundtland to head a new commission, to be called the World Commission on Environment and Development. Its mandate would be to investigate how "at a time of unprecedented growth in pressures on the global environment ... to build a future that is more prosperous, more just, and more secure because it rests on policies and practices that serve to expand and sustain the ecological basis of development".[7] The Brundtland Commission was to continue and to extend the work of Brandt's North-South Commission. It should integrate further the concepts of environment and development and formulate 'Earth politics' in the language of diplomacy to make it capable of securing international consensus.

Brundtland asked Brandt for advice. She had known him since childhood, from his time as an exile in Stockholm, where Brandt had been in contact with Brundtland's parents, like him exiles in Sweden. Willy Brandt offered his advice during a conversation in the margins of the celebrations for his 70th birthday in Bonn; he gave her the names of experienced people for the new Commission and tactical suggestions.

Gro Harlem Brundtland, a former national environment minister and Prime Minister, was a moderate Social Democrat. Full employment, consolidation of the welfare state, exploitation of the oil fields off the Norwegian coast while minimising risks and constructively critical loyalty to NATO were at the heart of her political beliefs. She was suspicious of political utopias. She was a skilled and accomplished practitioner of the art of the possible. This profile was what had convinced the UN functionaries to appoint her.

The new Commission met during the first days of October 1984 in Geneva. The Sudanese Mansour Khalid, whose country was at that very time suffering from a terrible famine, acted as Brundtland's deputy. Among the members were Maurice Strong, a key figure in 'Earth politics' since the Stockholm conference, the Italian Senator Susanna Agnelli, Saburo Okita, Japanese executive of the Club of Rome, William Ruckelshaus, former chief of the U.S. Environmental Protection Agency, the Columbian ecologist Margarita de Botero and the German ex-minister Volker Hauff. The Palais Wilson, an imposing nineteenth-century grand hotel on the promenade by Lake Geneva, now the seat of several UN institutions, became the headquarters of the Commission.

Thus began a two-year world tour of meetings, conferences and public hearings. The stops on this tour were Jakarta, Oslo, São Paulo, Vancouver, Nairobi, Moscow and Tokyo. The Commission had to grapple with the old frontlines which had been visible since Stockholm. They had become more deeply entrenched with the expansion of neoliberalism. Brundtland was aware that a repudiation of the dogma of growth could not be achieved against Thatcher and Reagan. She was probably convinced herself that growth was necessary in the countries of the North, too, to finance the transformation. Then again, for the representatives from the South it was a basic prerequisite that *development* and not the *environment* should be paramount. The additional costs of environmental protection should not be allowed to slow down the industrialisation of their countries, or to make it less affordable. Brazil refused to countenance any public discussion of the rainforests in the Amazon basin. Indonesia made sure that the public hearings were attended by a carefully selected audience. On the issue of nuclear energy, a public row was narrowly avoided. In the Commission, the Americans and Russians opposed the passionate demand for a complete withdrawal from the Columbian Margarita de Botero.

In December 1986, the negotiations over the report's conclusions reached their decisive phase during a meeting in Moscow. "We agreed", Gro Harlem Brundtland writes in her memoirs, "about how to describe the principal content of the concept 'sustainable development'."[8] The Commission went back to the formulations used in the World Conservation Strategy of 1980. Nitin Desai, an Indian economist who was one of the leading thinkers at the Geneva administration of the Commission, set to work with a small team on the editing of the final report. After four months the 400-page document, *Our Common Future*, was ready. On the 27th April 1987 the Brundtland Report was presented to the global public at a formal ceremony in London.

The key sentence in the original English text reads: "Sustainable development is development that meets the needs of the present without compromising the ability of future generations to meet their own needs."[9] This is the now famous and often-cited Brundtland definition. Its substance comes to life only in the context of the report. The report begins from the widest possible perspective – the planetary one. "In the middle of the 20th century, we saw our planet from space for the first time.... From space, we see a small and fragile ball dominated not by human activity and edifice but by a pattern of clouds, oceans, greenery, and soils."[10] This arresting image is followed by an analysis of a threatening crisis. "Humanity's inability to fit its doings into that pattern is changing planetary systems, fundamentally. Many such changes are accompanied by life-threatening hazards. This new reality, from which there is no escape, must be recognised – and managed." The root of the crisis is identified as contemporary human civilisation and its destructive impact on the planet's ecosystems. Humanity is confronted with the task of making the carrying capacity of the global ecosystems the measure of its actions.

The talk in this context is of *needs*. It is always *basic* or *essential needs* which are meant. So it is not 'need' in the sense of market 'demand' which is meant here, nor is it desires, and certainly not the assumed right to luxury of an elite upper class, but people's essential needs.

Development is then defined, at first, as enabling everyone to satisfy their essential material needs. Nothing more and nothing less. This is the primary aim of *growth*. In the report this term is open to different interpretations. There is no explicit rejection of the paradigm of economic growth. Nevertheless, the tendency is unmistakable: growth must take place primarily in those regions of the world where basic human needs are not yet being met. So growth means above all overcoming poverty. For the rich regions, the report considers the possibility of contraction processes and of strategies of self-denial. The Brundtland Report focuses not on continuing increases in the production and consumption of material goods, but on quality of life. This is where growth takes place: where it comes up against no limits. Because quality of life is measured principally at the level of immaterial needs.

'From one earth to one world' is the heading of the first chapter. The image of one Earth in space must become the political vision of one world, and thus of a 'global domestic policy'. This was a bold vision in the 1980s. At that time the world was still divided into the entrenched blocs of the First, Second and Third Worlds. Geopolitical conflicts were fraught with enormous risks. The danger of precipitating an all-out nuclear war, and of the 'nuclear winter' which would follow, was by no means over.

Sustainable development overcame the rigid thinking which dominated world politics. It was no longer a matter of 'conservation' and 'environmental protection' understood as a system of prohibitions, regulations and controls by means of which humankind was shielding the environment from the effects of its activities. Nor was it a matter of 'development' in the sense of a catch-up industrialisation of all of the regions of the world, nor of mere 'peaceful coexistence' between the blocs. *Sustainable development* is about the search for a new balance between humankind and nature, between the cultures of the world, and between people – it is about a new model of civilisaton. It would require a prompt and decisive mobilisation of all of the creative powers of humanity.

During the negotiations within the Commission in December 1986 in Moscow, as Gro Harlem Brundtland recalls in her memoirs, a performance of Tchaikovsky's ballet *The Sleeping Beauty* took place in the evening at the Bolshoi Theatre. This was exactly the time when the principal content of *sustainable development* was being hammered out in a marathon negotiating session. Brundtland was given a seat of honour in the Czar's box. This fitted perfectly: her Commission had just awoken the concept of sustainability from a long enchanted sleep.

In December 1989, the UN General Assembly decided to convene a global conference on environment and development. Its aim would be to convert 20 years' experience of 'Earth politics' into a model for the 21st century. The venue chosen was Brazil, the country in which some of the blue planet's most vital organs are located.

The spirit of Rio

The old conflict lines between North and South broke out again already in the run-up to the Earth Summit (1992), this time with more clarity and more acrimony. The mantra of the countries of the South was that the primary cause of the global crisis was the lifestyle of the North. More precisely, its dominant patterns of production and consumption. These have to be changed. The American President George H.W. Bush expressed the response with brutal clarity: the American way of life was not negotiable. The heart of the problem was now laid bare. It was looking as though the Earth Summit would fail.

The prelude to the conference was a gathering of representatives of indigenous peoples, from Brazil and from all over the world. In their traditional costumes, with their songs, dances and rituals, they celebrated and invoked *mater terra* – Mother Earth – in the streets of Rio, and they protested against the continuing plundering of the resources of their territories.

The end of the mammoth conference was marked by the signing of the various conventions and – by all the heads of state present – of an 'Earth Pledge', formulated and presented in Rio by Maurice Strong, Secretary-General of the Earth Summit: "I pledge to do everything in my power to make the Earth a secure and hospitable place for present and future generations." 108 signatures were gathered, including those of Gro Harlem Brundtland, Fidel Castro and even George H.W. Bush. In all the years of Earth politics, solemn oaths to save the planet have never been in short supply.

The conference itself however was anything but a harmonious exchange of views under a clear tropical sky; there was no happy meeting between Riocentro, the monumental concrete congress centre, and Flamengo Park, far across town near the beach, the venue for the Global Forum of the NGOs. Disaster threatened more than once. Substantial declarations fell victim to verbal sparring matches, sometimes heated, sometimes conducted with cutting coolness. The conventions on climate protection and on biodiversity, and the action programme for Agenda 21, all of which had been carefully prepared in the UN over several years, were diluted. An outline for an 'Earth Charter' was replaced by a brief declaration of its principles. At Rio the conflict lines surfaced for everyone to see: North against South, governments against NGOs, the USA against the rest of the world. The USA – and, hidden in its wind-shadow, many other countries of the North – fought hard to protect free markets against any form of regulation. The South defended with bitter determination its sovereign right to exploit its own resources. The

North, it was argued, was morally obliged to provide the countries of the South with new and additional financial resources and with technical know-how for the protection of the environment. The NGOs of North and South took up radical ecological and social positions. Déjà vu: the frontlines from the Stockholm conference of 1972 were here again. And at the UN Rio + 20 conference in the summer of 2012 it was truly crushing to see that they had still hardly moved.

Nevertheless, the epochal event of the Earth Summit was the establishment of sustainable development as the global guiding principle for the 21st century. "Recognising the integral and interdependent nature of the Earth, our home", reads Article 1 of the 'Rio Declaration', the Conference proclaims that: "Human beings are at the centre of concerns for sustainable development. They are entitled to a healthy and productive life in harmony with nature ... The right to development must be fulfilled so as to equitably meet developmental and environmental needs of present and future generations." In this consensus lies the importance of the Earth Summit. It was the step up to the next level.

Sustainability is an ethical principle. The foresters of past eras knew this, and so too did the pioneers of Earth politics. The Brundtland Report too contained a plea to elevate sustainable development to a global ethic. In the run-up to Rio, Maurice Strong and his team had sketched the outlines of an Earth Charter which would give an ethical foundation to Agenda 21. But he had to acknowledge during the preparations for the conference already that it would not be possible to achieve a consensus around a document of that kind. Following Rio, Strong and his colleagues set about closing this gap. He was able to find partners in civil society movements from around the globe. In addition to numerous grassroots intitiatives, Wangari Maathai, the Kenyan who had started a tree-planting campaign among the ordinary people of her country, Steven Rockefeller, an American philanthropist, Mikhail Gorbachev and the Argentinian singer Mercedes Sosa all engaged in the project. The UN university in Costa Rica became its hub. After a process of debate and consultation lasting several years, the Earth Charter was presented to the global public in the spring of 2000. Since then the UN and UNESCO and representatives of several governments have taken favourable note of the Charter. But no more than that. And yet it holds "the best-thought-out, most universal and most elegant blueprint for a global ethic there has yet been", according to the South American liberation theologian Leonardo Boff. Moreover, the Earth Charter is the boldest formulation to date of the principle of sustainability. It overcomes the anthropocentric limits of the idea without sacrificing its humanist core.

Its first principle: "Respect Earth and life in all its diversity. Recognise that all beings are interdependent and every form of life has value regardless of its worth to human beings." [11]

The conclusion which follows from this: "We must join together to bring forth a sustainable global society founded on respect for nature, universal human rights, economic justice, and a culture of peace."

The closing sentence of the Earth Charter: "Let ours be a time remembered for the awakening of a new reverence for life, the firm resolve to achieve sustainability, the quickening of the struggle for justice and peace, and the joyful celebration of life."

That every form of life has intrinsic value. This is the hub around which the Earth Charter revolves. It is the starting-point of the rationale for sustainability. In this, the Charter is basing itself on the ethic of reverence for life, and therefore on Albert Schweitzer. The phrase refers to an experience which the great humanitarian had in 1915. During a boat trip on the Ogowe river in the African tropical rainforest the words 'reverence for life' came to him at the sight of a migrating herd of hippopotami. Schweitzer wrote in 1923 that the phrase expresses a philosophical position "which places Being above Nothingness, and therefore endorses life as something intrinsically valuable". This holds the essence of sustainability. It is a way of thinking which is born of crisis. But in the midst of crisis it draws together the energies which spring from the affirmation of the world and of life.

A circle is nearing completion: *abad* and *shamar* – to tend and to keep – to conserve and to utilise – sustain and develop; the Franciscan communion with nature, Spinoza's ideal of the *multitudo* in harmony with the *natura naturans*, Goethe's dream of the rejoicing universe, Albert Schweitzer's reverence for life – the ancient visions combine with the latest insights of Earth politics and reappear in the international arena – forever young. Something came into the world which had far-reaching implications. We still haven't understood the full extent of its reach. The discovery of sustainability continues.

Rio + 20, the UN conference commemorating the 20th anniversary of the Rio Earth summit, took place at Rio de Janeiro in June 2012. "We... renew our commitment to sustainable development" was the first sentence of the final declaration entitled "The world we want". Apart from that the event turned out to be a summit of vagueness. Some observers even called it an utter failure. Did the summit reflect a – temporary – weakness of Earth politics as defined 20 years previously at the same place? And a – just as temporary – worldwide regression to the old geopolitical power games? Those are definitely not the "future we want", nor what the planet can tolerate.

CHAPTER FOURTEEN

What next?

It is time to return to where we started. How do we distinguish between what is sustainable and what is unsustainable? My personal litmus test has two components:

1. Does it reduce the ecological footprint?
2. Does it widen access to a good quality of life?

For something to be 'sustainable', it has to pass these two tests. But why precisely these two?

The first test is a response to the current dramatic over-exploitation of the ecosystems. Each year in my diary I highlight a special date in red. In 2011 it fell on the 27th September. On this day in early autumn, the stock of natural resources produced by the planet over the entire year 2011 had been used up. On this date, the quantity of waste dumped and emissions blown into the air reached the limits which can be absorbed in one year by the ecosystems. American ecologists call this day 'Earth Overshoot Day'. Since 1986 we have been 'overshooting' this invisible line every year – and every year we have done so a day or two earlier. The forces pushing the Earth towards a future of turbulence remain stronger than the countervailing forces. They are carrying us further away from the goal of sustainability. After four decades of Earth politics the planet remains on course for the collapse which the Club of Rome forecast would happen around the middle of the 21st century.

Reducing our ecological footprint means synchronising our lifestyle and our economic cycles with natural processes once again. To say it in the language of Linnaeus: to embed the human economy (*Oeconomia nostra*) once more in the economy of nature (*Oeconomia naturae*). What we need above all is the courage to reduce, to recognise that 'small is beautiful'. Sustainability was and is primarily a strategy of self-control and of reduction. We will save the planet by using its ecosystems less, with more care, and in a different way.

From the *Ur*-texts right up until today, sustainability has always had a second constant: the search for *summum bonum*, the highest good (Seneca), for *beatitudo*, the happiness and blessedness which Spinoza, Leibniz and their contemporaries had in mind, for the joyful celebration of life which the Earth Charter proclaimed. The pursuit of a higher quality of life clearly has something to do with growing, with advancing and with achieving fullness. Without growth, human existence is not a true life, and not worth living. But what is meant by growth here is something qualitatively different from the accumulation of goods and the raising of GDP. This is about personal and collective growth, about freedom of access for everyone to the higher levels of the pyramid of needs, about enjoying diversity and richness in nature and in the cultures of the world. The promise of the good life for all draws us into this open realm where there can be no 'limits to growth'.

This dimension of sustainability, too, is more topical than ever. In the era of turbo-capitalism, our resources of community spirit, trust and solidarity melted as fast as the ice on the polar glaciers. But a confident expectation of a high quality of life is the best possible guarantee of social cohesion. "There is enough in the world for everyone's needs, but not for everyone's greed." Since the collapse of the banks, this saying of Mahatma Gandhi's has been popping up on countless websites. The password here is 'enough'. Being content with enough – or better, knowing what is enough – is the essential virtue of the art of living sustainably. In the jargon of the specialists, it is called sufficiency. The sustainable society will be more egalitarian, and more just – or it will remain a dream.

I prick up my ears when sustainability or one of its associated terms fails to appear in any kind of programme or master plan. Then I pay close attention to the sources from which the key terms are drawn. Usually, the vocabulary of the globalised economy is at the forefront: free markets, growth, progress, competitiveness, purchasing power. This terminology already seemed to have been rendered obsolete 40 years ago, when the 'limits to growth' entered the public consciousness. The fact that it was soon quietly rehabilitated and then allowed to spread like never before, and to colonise new areas of life, was a disaster. Today, in the looming shadow of climate catastrophe, this vocabulary is an inherited semantic burden, just as toxic as the legacy of industrial garbage and hazardous waste sites, or the securities in the vaults of the 'bad banks'.

Neither Linnaeus nor Goethe would have been able to comprehend the fetishisation of what we call 'growth'. In their time, this word referred in a straightforward way to natural processes, namely the biological growth of trees, of corn, of animals. Not until the fossil-fuel age did this semantic shift occur. Coal and oil do not grow. They lie in store. Stores, however, can be opened up at any time. More and more previously untouched deposits became available, apparently limitless in number. Thus was the extraction of resources, the production of goods, and – the ultimate goal – the increase of invested capital continually raised and accelerated. Only then did 'growth' become a religion. The Swedish writer Kerstin Ekman has pulled the veil off the fetish doll. "We worship a principle, only one. It has taken its name from the growing woods, but it makes nothing grow. It is called growth, and what it increases is artefacts." The unchecked proliferation of artefacts is precisely the problem. Their quantity exceeds any conceivable capacity for ultimate disposal. A plastic bag takes between 100 and 400 years to decompose, a burnt-out fuel rod from a nuclear power station half a million years – and more. The acceleration of growth ends with a pillaged, bare, contaminated planet.

Against this background, discussion of the three pillars or the triple bottom line of sustainability appears in a new light. This discourse was originally intended to underline the interconnectedness of the ecological, economic and social dimensions. In fact, though, its aim is often turned on its head, and it is now used to suggest that the ideology of growth and the philosophy of sustainability can co-exist.

It is claimed that the three dimensions – or pillars – of sustainability are of equal importance. 'People, planet, profits' is then invoked to suggest that we can only look after the environment and our social objectives if the economy grows. Growth pays for sustainability. A variant of this is to define sustainability only as a 'market segment'. There is of course nothing wrong with making money from green products and services. But it would be a serious error to try to turn them into the new engine of growth and to block off other pathways to a sustainable development. Viewing the economy through the prism of sustainability leads to a different conclusion: the necessity of breaking the hegemony of the cost-benefit analysis – the 'terror of the economy' – and speeding up the transformation into an ecologically sound moral economy.

At this point the purely managerial approach loses its stranglehold and makes way for a philosophy of 'let it be'. Not in the sense of 'laissez-faire' – letting everything happen without regulation. But in the sense of letting go – leaving alone, or leaving intact. Not doing something even though it lies within one's power. This kind of leaving or letting belongs to the essence of sustainability. It requires not only courage but wisdom.

Dejar el petróleo baja tierra! 'Leave the oil in the soil!' Under this slogan, a grassroots movement in Ecuador is trying to preserve the Yasuní territory in the Amazon Basin. The area is a treasure house of biodiversity, the home of an indigenous people – and the site of a huge oil deposit. To forego the presumed benefits of exploitation, to prevent contamination, to protect the climate, to leave the integrity of land and people intact – that is the high road. Whoever takes it deserves every support.

Let it be – to forego something that is within one's power – that wisdom is inscribed in the concept of sustainability. That is its great strength.

Of course, the vocabulary of the age of fossil fuels doesn't derive its appeal from the imagery of the brutal suppression of nature. What is it, in fact, that gives it its glamour? Beneath the surface, an ancient symbol is working its magic on us, one which originated in the classical world, enjoyed a new lease of life in the Renaissance and is today celebrating a comeback, often without being represented by a visual image. It is the the horn of plenty. It promises an eternal superabundance of material goods and unlimited access to all kinds of pleasures. This promise has deep roots in the world's cultures.

> Six kings were dethroned
> four seas were united
> the mountain of Shu was chopped off
> and then E Fang Palace was built

The first lines of the Chinese poet Du Mu's 'Ode to E Fang Palace', written in the 9th century, reveal the dialectic of natural destruction and – apparent – cultural flourishing. A transposition to today might read: The globalisation of the markets completed, any spot on the globe reachable within 24 hours, peak oil past, Dubai megacity was built – and the International Space Station (ISS).

Creating artificial Edens requires human interference with nature on a massive scale. We turn a blind eye to it, or we think it a price worth paying, in the belief that we will eventually be able to completely detach ourselves from nature and make our home in a world of artefacts. The ecological crisis is an alarm bell warning of the end of nature. But what does that matter to us? Technological progress is feeding our fantasies of omnipotence, of a *conservatio sui* – of human self-preservation – by means of technology, without nature. The Earth itself will gradually be transformed into a gigantic artefact, and thus into an unlimited source of luxury and of the good life – an artificial horn of plenty. Yet every square centimetre of aluminium on the outer skin of the ISS, every litre of water which emerges from the showers of the hotels of Dubai, is a gift of nature. An entirely deforested, desertified, despoiled Earth will no longer provide coming generations with the basic essentials of life. So does the symbol of the horn of plenty belong to the toxic semantic legacy which we need to leave behind us in the 21st century? When used in this technophilic sense – yes! But the cornucopia

190

of classical mythology bestowed a gentle flow of nectar and ambrosia – gifts of nature, easily renewable resources.

In the first century BC, the Chinese philosopher Dong Zhongshu celebrated human civilisation as a precious part of Earth's unique community of life. "The influence of *tian*, of the interplay of heaven / cosmos and earth / nature, ends with the silk cocoon, with the hemp plant and with the straw. But to make cloth out of hemp, silk from a cocoon, a meal out of rice grains … all those things can be achieved only when wise humans continue and transcend nature's actions. None of them can be achieved by nature on its own." The message for today's world is that ecology does not mean a regression to a state of nature, and that artefacts are not necessarily destructive. When they are successful they add something to nature. To turn the scarcity of resources into a stimulus to innovation, to create the maximum wellbeing from the minimum resource – that would be the contemporary symbolic role for the horn of plenty.

Climate change has brought forth some remarkable semantic innovations. Their effect is always to concentrate and focus things onto one point: CO_2. We hear of the dawn of a climate-neutral era, or carbon-free development, or a low-carbon society. No doubt what is happening here constitutes an important supplement to the language of sustainability. However, it would be a disastrous mistake if a discourse based exclusively on climate change were to displace the sustainability discourse and crowd out all other issues involved.

The beginnings of just such a trend can be traced back to the summer of 2005. At the end of August, Hurricane Katrina swept over the Caribbean. On the third day, the New Orleans levees were breached and masses of water from the Gulf of Mexico flooded 80% of the residential areas of the city. For the inhabitants a tragedy; from a global perspective, a catastrophe of middling severity. But the overnight collapse of a major First World city came to be seen as the writing on the wall. Climate change, the very existence of which had previously been the subject of bitter dispute, was suddenly front-page news everywhere. The warnings which had been sounded for decades against climate change as "potentially the most serious catastrophe in the history of human civilisation" (Al Gore) were now taken seriously.

The British astronomer Martin Rees, President of the Royal Society – at whose gatherings 350 years ago John Evelyn had warned about the side-effects of burning 'sea-coal' – published a book with the title *Our Final Hour*. He warned that the chances of our civilisation surviving until the end of the century stood at 50:50. Other experts warned of resource wars and of a 'long emergency'.

The wind had turned. Finally, climate change was no longer being denied, nor its man-made nature. But rather than being trivialised, its danger was now exaggerated. Worst-case scenarios flooded the media. Rising sea-levels, the diversion of the Gulf Stream, heatwaves, droughts, conflagrations, epidemics, climate wars – the tone grew shrill. Images of biblical apocalypse were dug out again. Hell on Earth drew nigh. As real as the danger of climate catastrophe is, this kind of language can do as much harm as good.

More useful than a proliferation of Armageddon scenarios is a reminder about the angry warnings given by the countries of the South at every major UN conference from Stockholm via Rio to Copenhagen: the major cause of the global crisis is the model of civilisation adopted in the North – more precisely, the *patterns of production and consumption*. These must be altered. The deeply inconvenient truth is that these warnings were right. That is the core of the problem. And they are still right.

"The point is that climate change is not 'the problem'. The problem is a sprawling civilisation that is destroying the biosphere," the American ecologist Eileen Crist wrote in a remarkable essay in 2007.[1] Climate change is only the most pressing current symptom of an enveloping crisis whose ultimate cause is the industrial-consumer civilisation. Crist bases her analysis explicitly on Herbert Marcuse's critical theory.

The fixation on climate change blinkers us to possible solutions. The resulting tunnel vision then only allows us to see the technical fixes offered by the industrial system. Instead of sustainable development, now the talk is of Plan B. Where until recently the argument was heard that climate change had not been conclusively demonstrated, where the word *sustainability* was met with a disdainful and irritated rolling of the eyes, now you hear that all attempts to prevent climate change are doomed to fail and that the consequences are unavoidable. Now everything has to be focused on managing adaptation to those consequences. This argument completes the rehabilitation of the technological approach, from small-scale mitigation projects right through to fantasies of geoengineering: the monocultural cultivation of energy crops, CCS (carbon capture and storage), nuclear energy, nanotechnology, and – not least – increased military expenditure. Climate change policy, though, is not a new market opportunity but a survival strategy. Apocalyptic predictions are seldom helpful in this context. They pave the way for authoritarian responses.

Eileen Crist points out the dead end we have stumbled into through our failure to address the deeper causes of the ecological crisis. "The destructive patterns of production, trade, extraction, land-use, waste proliferation, and consumption, coupled with population growth, would go unchallenged, continuing to run down the integrity, beauty, and biological richness of the Earth."[2] Even if we were to succeed in 'managing' climate change (keeping the temperature increase within the upper limit of two degrees centigrade), we would not stop the continuing destruction of life on Earth. This is about more than going 'carbon neutral', more than the transition to new methods of resource management; it is about a *great transformation*.

In this world-wide movement for the great transformation, new ideas, images and catchphrases are emerging. They address the dialectics of material reduction and spiritual growth. They are no longer conceived as rigid containing or guard rails, but rather as navigational landmarks on an unknown but excitingly beautiful landscape. They relate to a variety of needs, to liveability and attractiveness, to sensuality and aesthetics; in short, to the good life, both in the present and in a desirable and feasible future.

Reduce, reuse, recycle is the title of a little ditty performed by the Hawaiian surf icon and pop star Jack Jackson in the kindergartens of his home state, and sent around the world on video. The technical-ecological principle of the waste hierarchy is turning into a module for pre-school learning and moreover into a natural daily routine to free up your head while waiting for the perfect wave to come along. Voluntary simplicity, as a less-is-more philosophy, is discussed on Oprah Winfrey's website in the USA. It's all about "trading possession obsession for personal fulfilment". 'Simple in means, rich in ends' is the maxim coined by the Norwegian philosopher Arne Naess, who died in 2009. 'Nothing in excess' – one of the inscriptions from the Oracle of Apollo at Delphi – has been recycled as a motto by a German climate change organisation. The credo of former Bogotà mayor Enrique Penalosa is: "A city is successful not when it's rich but when its people are happy." It could just as well serve for the Transition Towns movement which originated in the Irish town of Kinsale and the British town of Totnes around the year 2005. With a close connection to the ideas of permaculture gardening, it aims at raising awareness for sustainable lifestyles and ecological resilience. The Transition Towns movement is a rapidly expanding network of communities with affiliations in Australia, North America and Europe. The Himalayan kingdom of Bhutan, to quote just one more example, no longer measures its success by changes in Gross Domestic Product, but by means of a Gross National Happiness Index. And so the design of sustainable lifestyles progresses via separate but networked movements around the world. The simple life can be the good life.

'Slower – less – better – finer'. These coordinates for a new vision of the good life were devised by Hans Glauber, an Italian artist, sociologist and mountain climber in 1992, the year of the Earth Summit, and he continued to develop them together with a small, fluctuating circle of maverick thinkers in the 'Toblach Talks' until his death in 2008. Over that time, the focus shifted from the technical implementation of ecological transformation to the cultural dimension of sustainability. Notwithstanding its criticism of the destructive tendencies of modern civilisation, this initiative took very seriously the temptation of 'faster, higher, stronger' – and 'more'. It did not underestimate how difficult it would be to displace the growth paradigm and its promise of stability and security. Nor did it fail to acknowledge the aesthetic fascination and the promise of happiness which the world of consumption holds. In an open competition between the two systems, the new model would need to increase its appeal and to demonstrate that it was superior.

The basic strategy was developed step by step in the recurring discussion rounds. The solar era meant essentially doing what humankind had always done. The fossil-nuclear era is only a short episode in history; it stretched from the beginnings of industrialisation until – at the very latest – the exhaustion of the fossil fuel deposits. Before that humankind had lived from the Sun alone. Afterwards it would live from the Sun again. But life in the second solar era could be organised in a much more civilised way. Thanks to new technologies – above all, the possibility to create electricity from sunshine – we will be able to exploit solar energy much more efficiently and flexibly.

The new model will require comprehensive reform of our way of living together. It will be a more decentralised, more democratic and more equal civilisation. For in contrast with oil, and with the other fossil and nuclear fuels, ownership of which is concentrated in very few hands, the Sun shines for everyone. We don't own the Sun; we only have access. Moreover, as a source of energy, sunshine has the advantage that it is particularly plentiful precisely where poverty today is most widespread. The utopia of a more just development is coming within our reach. The new model of civilisation is posited on a new balance of material and immaterial goods, on quality of life rather than on an unbalanced wealth measured only in goods.

"The economy of the good life is made up of an environmentally-friendly combination of measured consumption and the enjoyment of immaterial goods." (Toblach Theses) The solar era makes possible a civilisation which is substantially less resource-intensive. It is based on new ways of producing and consuming. It acknowledges the necessity of living creatively within limits. It turns the inescapable fact of quantitative limits into the basis for a different approach and seeks the potential for vibrant continual growth in the sphere of immaterial goods and values. The function of material goods is essentially to make it easier for us to produce immaterial and common goods. Limits represent an opportunity. They themselves become an additional resource. What we have to do is to get the maximum output while remaining within these limits. This results in things which require very little material and which thus develop their own very specific style. An aesthetics of moderation is emerging out of this.

Beauty, too, is a kind of nourishment: like food, it is essential to life – a victual. Without beauty no life can be fulfilled or fulfilling. Active engagement in the movement for a sustainable culture often springs from the experience of damaged beauty – for example, from disfigured landscapes, or from urban dreariness, in the wake of industrial mass production. Such a culture must speak to the

194

sense of beauty. The beauty of moderation unfolds in the caring and careful use of resources. It emphasises local character and tradition as much as natural and cultural diversity. The enjoyment of organic foods, the sensuous experience of nature, the pleasure to be had from good design and good architecture all contribute to the joy of life. "Beauty is truth, truth beauty"(Keats). It is also one of the strongest drivers of human actions, and must be included in comprehensive ecological transformation. Beauty is one of the building materials of our common future.

My prediction is that 'sustainability' will remain the key term. It has the necessary gravity and the necessary flexibility. This word contains everything that matters. As Hans Glauber said, "The vision is the solar era, the era of an all-encompassing new culture of sustainability."

The word derives its gravity from its existential perspective. Its subject from the very beginning is the human capacity to look ahead into the future and to provide for the chain of coming generations. Its horizon is therefore the entirety of space and time – infinity. Ecology and the quality of life, including global justice, are captured and stored at the core of this concept. Without ecological stability and social cohesion there will be no future worth living.

What gives the term its flexibility? It is its capacity to adapt its meaning to the prevailing conditions. Sustainability, I learned in an interview with the nuclear physicist Hans-Peter Dürr, is not a rigid system of rules, not a precept which tells us what to do in any possible situation. "Ultimately, what it does is to give us a realm in which to experiment without the risk of total failure, room to create new combinations in order to learn from them and to make further new combinations. Sustainability is desirable in the sense that we don't want to be eliminated from the evolution of living things."

This plea can be found already in the Brundtland Report: keep the options open. Looking ahead and providing for the future certainly doesn't mean prescribing how coming generations must live, but rather keeping the options open so that they may be able to fulfil themselves in line with their needs and wishes. And this requires that today we preserve biodiversity and cultural diversity.

The gravity and flexibility of the term sustainability have enabled the growth of a powerful and continually self-renewing semantic force field. The 'energeia' (Humboldt) of the main term draws in semantically-related words and influences them. And this results in reciprocal interaction which re-energises the main term. Right now it seems to me that the word field around sustainability is expanding. Everywhere in the world it is stepping over into everyday usage, where it acquires associations with sensory experience and thus a sensual quality.

The acronym LOHAS (lifestyle of health and sustainability) couples *sustainable* with *healthy*. Whatever precisely may be understood by that, this verbal alchemy dissolves fixed meanings and vitalises old words. In this way, too, counter-terms to the 'growth' paradigm acquire greater appeal; 'throwing ballast overboard', or 'downsizing', is certainly conducive to health. It makes us – another related word – *resilient*, capable of maintaining our selves and our dignity.

The word *green* is becoming fashionable all over the world. "The world is fortunately beginning to turn green", said the biologist Edward O. Wilson in a speech in 2011, "at least pastel green". A *green revolution* is said to be under way. It is in this context that phrases like *low-carbon society* acquire momentum.

The joining together of *sustainable* and *development* is not without its problems. Where 'development' is understood only as 'commercial opportunity' and 'economic growth', this pairing serves to increase conceptual confusion. But where development is understood in its original sense, as the unfolding of potential, there it serves to add dynamism to the concept of sustainability.

We are living in a time of change. What should we call this process? The age-old term *transformation* is today increasingly associated with the semantic field around sustainability. 'Resist, mobilise, transform' was a slogan used by some of the NGOs at the Copenhagen Climate Conference. More conferences are taking up the theme of 'the great transformation'. The word indicates the dimensions of the changes needed and makes it possible to envisage the total conversion of structures. The word was familiar to Francis of Assisi and to the mystics of the Middle Ages. Today it is more comprehensible than ever. Every electrician knows what it means.

One world – *unus mundus*. In many cultures of the world attempts are being made at the moment to integrate *sustainability* with national and local traditions. *Sumak kawsay* is the term for the good life in Quechua, the language of the Incas. Today under this heading a discussion is going on about a life in harmony with *pachamama*, Mother Nature. The oldest and most inclusive concept of Chinese culture is the subject of a revival of interest just now. The philosophy of *Tian-ren-he-yi* has deep roots in the Confucian, Taoist and Buddhist traditions. It teaches that Heaven/Nature and Mankind/Society are interconnected. In China *xietiao fazhan* (harmonious development) is on everyone's tongue. Is it just an empty slogan? Or is it the organic combination of the old philosophy with the modern discourse of sustainability? In India, Mahatma Gandhi's 'trinity' of concepts is experiencing a revival in connection with sustainable development. As described by the British author Satish Kumar in his book *Spiritual Compass*, they are, firstly, *Sarvodaya* – the *upliftment of all*, or *All rise*. Meaning not just a few, nor even only a majority; *Sarvodaya* includes the interests of all people as well as care for the Earth, animals, forests, rivers and land. The second principle is *Swaraj* – self-government, through small-scale, decentralised, self-organised and self-directed structures. And the third is *Swadeshi*, the revival of the local economy and of manual work, of manufacturing in small workshops and the practice of arts and crafts that feed the body as well as the soul.

José Lutzenberger, the Brazilian environmentalist and trailblazer for the 1992 Rio Earth summit, once expressed his hopes like this. "If we can arrive at an ethic of reverence not only for life but for the cosmos itself, then we can build a fantastic civilisation." The front is long. The future is an untrodden path. Even if the window of opportunity is beginning to close – I place my trust in swarm intelligence.

Epilogue

A stone's throw from Goethe's garden house in the park by the river Ilm in Weimar there stands, next to a fallen linden tree at least 200 years old, at the end of a path lined by hollyhocks leading to the outer fence, a shoulder-high sculpture made of sandstone. Goethe designed it himself and had it placed at this spot in 1777, shortly after moving into the garden house. The 'Stone of Good Fortune' is remarkably plain and simple in form: two pure geometrical shapes, a cube about 3 feet by 3 feet supporting a sphere of slightly smaller dimensions, both made of the same stone. The centre of the sphere is exactly in line with the centre of the cube. Its curve only lightly touches the flat surface. One almost has the impression that the sphere is floating above the cube. Two basic geometric forms set in a small, fenced-in nature park. Nothing more.

This is a place of tranquillity. But if you sit down for a while on the bench or on the fallen tree trunk, the tranquillity reveals itself as a sound sculpture made of delicate noises. Depending on the time of day and year you can hear birdsong, the humming of insects, the rustling or sighing of the leaves in the wind, occasionally the voices of visitors to the garden house, or, from the distance, very faint, the sound of the castle bell – just as it sounded in Goethe's time. The fallen winter linden tree at one time stood 20 metres high. Perhaps it cast its shadow on the stone already in Goethe's day. A storm brought it down at the end of the 1960s. Since then, a few young trees have grown upwards from the many sleeping (or 'adventitious') buds along the trunk. These are coppice shoots.

The garden exudes the smell and sense of the soil. One's thoughts journey backwards. This is where Goethe enjoyed the rays of the early spring Sun, growing stronger by the day – and the energy of the Moon. "Flooding with a brilliant mist / Valley, bush and tree, / You release me, oh for once / Heart and soul I'm free." On this spot he harvested asparagus and strawberries. In this environment he conceived his theory of needs and lived accordingly. "I had to endure absolutely horrible weather today. . . Almost everything is soaked. . . I'm now drying out my things! – they're hanging around the stove. How *little* one needs, and how good one feels when one realises how *much* one needs that *little*." And Herder, his "dear brother", concurs: one has to "enjoy ... every moment, and to desire nothing but what Nature offers us." Does that mean deprivation? "Doing without," Herder says, means for him only "removing from my life what doesn't belong there."

The 'Stone of Good Fortune' rests with all its weight on the earth. Its cube expresses groundedness, steadfastness, gravity. The sphere by contrast symbolises the element of mobility, the momentum of the rolling stone, the force of perpetual change. Cube and sphere, the fixed and the mutable, are complementary. From the synthesis of gravity and flexibility a new entity emerges. This is a flexible order. Each time I see Goethe's 'Stone of Good Fortune' I experience it anew, as if for the first time, as a symbol of sustainable development.

"La rivoluzione siamo Noi" (We are the revolution), Joseph Beuys wrote on a photo of himself. It shows him in the last year of his life, as a hiker, with a light pack. 'Capri Battery' is a small art work from the same period. The approximate value of the materials involved is today about one Euro. A light bulb sticks up at an angle. The glass of the bulb is a luminous yellow. The base of the black plastic socket is plugged into a lemon, just as an ordinary light bulb is plugged into an energy source. The two elements, light fitting and lemon, support each other in a delicate equilibrium. The lemon prevents the light fitting from falling over, and the fitting holds the lemon steady. That's all. Minimal means, simple materials – *arte povera*.

"Joseph Beuys, Capri Battery, change battery after 1,000 hours" – that was the inscription on the multiple, created in 1985. It is a paradoxical image of energy delivery.

An electric circuit is completed. The solar energy which has been transformed by the plant, Capri Battery suggests, flows through the light bulb and makes it shine continuously. The sunny yellow colour and the sun-like forms of the fruit and the artefact refer to the cosmic power station that is the Sun and to the elemental cycle of energy: here the Sun, there the artificial light. The laconic instruction manual draws attention to the finite but renewable nature of the energy storage system. Once the juice of the lemon (the electrolyte) has dried out, the fruit (the battery) has to be changed. It is of course possible to do this. The bioelectricity produced by the fruit is very weak, but it is real. And it is renewable. It is created from a raw material which replenishes itself.

In the second decade of the 21st century, there are clear signs of the emergence of powerful new narratives of sustainability, of a new sustainable aesthetics, and of a green (or green-ish) popular culture.

In December 2009, as the UN Climate Change summit in Copenhagen was failing, with potentially catastrophic consequences, a Hollywood blockbuster had its world premiere. Within a few weeks it had become the biggest box office success in film history. *Avatar* is set in the year 2154. Planet Earth has been stripped bare. The search for new resources has shifted to distant planets – but using the same old methods. The mining company RDA plans to exploit deposits of an extremely valuable metal called 'unobtanium'. These, however, are on the planet Pandora, in an enchanted rainforest under the 'Hometree', a sacred tree that grows on the land of an indigenous tribe who live a simple life close to nature. The tribespeople defend their territory against colonisation. They take up arms in an unequal conflict. The invaders bring in heavy equipment: space shuttles, battle helicopters. They march in military order across unknown territory. Of course, they are headed for defeat. The native people roam easily on foot through the deep jungle of their planet with elegant, smooth movements, or else they ride on beautiful mysterious creatures. Jake Sully, an earthling, becomes fascinated by this strange world and by Neytiri, a female member of the tribe. Where is it that they finally get close to each other? On a night walk, lit by shimmering phosphorescent plants, to a waterfall on Mons Veritatis, the mountain of truth.

Just a silly fantasy film? *Avatar* tells the story of a clash of cultures that might turn out to be the real one, the one that determines our future. Leave the oil in the soil, as the indigenous people of Yasuní National Park in Ecuador say. Let it be as it is. Let nature be nature.

A few weeks after the premiere of *Avatar*, the simple bronze figure of a walking man became the most expensive object in the history of the art trade. 'L'homme qui marche', a masterpiece by the Swiss sculptor Alberto Giacometti, was sold by Sotheby's in London for the world record price of $104 million. Why did this statue, created in 1960, apparently enjoy such a sudden leap in value in 2010 (the anticipated sale price had been $20 million)? It depicts nothing more than the silhouette of a man taking a step forward. The sculpture captures the walking man in the fraction of a second when his back foot lifts off from the ground, while the other foot swings out to the front and comes down. The man's body is slightly bent at the hip. From his centre of gravity he draws the energy for the next space-consuming step. The pedestrian is not a muscular athlete, but rather a spindly, one might even say 'wiry' figure. The accentuation of his long limbs reveals the flow of energy that pulsates through everybody when they walk. Giacometti's sculpture embodies the dynamics of self-propelled motion, the dignity of walking upright – an ancient metaphor for freedom – and the power of resilience.

Is that its secret? In an age and in a civilisation where we spend up to 90 percent of our lives in enclosed spaces – increasingly in the even more enclosed virtual worlds of cyberspace – and see the world almost exclusively through windshields and on monitors, this work of art sends us a simple message: you are a creature in motion, and a creature of movement. Recognise your freedom to break away and go wherever you will. Just set off.

And listen to what the Indian writer and activist Arundhati Roy said at the assembly of the World Social Forum in the Brazilian city of Porto Allegre: "Another world is not only possible, she is on her way. On a quiet day I can hear her breathing."

Last year, in a small town in Germany, the women of the local farmers' association celebrated its anniversary by planting fruit trees in a school garden. When asked by the local newspaper why they had chosen to mark the event in this way, a spokeswoman said, "We just wanted to do something sustainable."

Acknowledgements

This book grew out of countless inspiring conversations and extensive correspondence with experts in different fields. It grew out of years of interdisciplinary reading, and – last but not least – from travelling to places which are imbued with the history of sustainability.

I am deeply indebted to many people for their input which has helped me in vital ways, and I would like to thank them all. My gratitude for setting the English edition on its way is due in particular to Klaus Bosselmann and Prue Taylor, to Jorgen Randers and Udo E. Simonis, to Ernst Ulrich Köpf and Donald Worster, to Ernst Basler and Hermann Graf Hatzfeldt, and to Susanne Eversmann. I would like to thank them all for their encouragement, advice and support. My heartfelt thanks also go to Ray Cunningham, who did a wonderful job translating the book from German into English. In fact, his expertise in both languages and traditions as well as his knowledge about the current discourse on sustainability helped to update and improve the book.

References

Chapter One
1. Goethe (1989), p.303.
2. Whitney (1905), p.667.
3. HRH The Prince of Wales. Foreword to Pye-Smith (1994).
4. Meadows (1972), p.158.
5. Christopher G. Weeramantry, Foreword to Bosselmann (2008), p.vii.

Chapter Two
1. WCED (1987), p.43.

Chapter Three
1. Archibald MacLeish, quoted from Poole (2008), p.8.
2. For this and the following quotations see the Apollo 17 Flight Journal (1972). Available online at www.ehartwell.com.
3. This and the following quotations from the astronauts' discourse come from Kelley (1988).
4. See Freud (1961), p.11
5. World Commission (1987), p.1.
6. Carson (2000), p.22.
7. The story of the Boulder conference is told by Weart (2008), pp.38-62.
8. See Maslow (1987).
9. WCED (1987), p.8.

Chapter Four
1. See Leclerc (1977)
2. Carson (2000), dedication to Albert Schweitzer.
3. Meditations of Marcus Aurelius, chapters 2 and 7. Available at www.felix.org.
4. The first epistle of Clement to the Corinthians, transl. by J. B. Lightfoot. Available at www.earlychristianwritings.com.
5. This and the following quotations are from Niavis (1953). See also Merchant (1980) Chapter 1, and Adams (1938), pp.170-190.

Chapter Five
1. Rifkin (2004).
2. See Cusa (1981).
3. See Kepler (1967).
4. Kittler (2001), S. 37.
5. Descartes (1998), p.35.
6. See Spinoza (1883). Available at different websites, e.g. www.gutenberg.org.

Chapter Six
1. For an in-depth analysis of the Venetian sources see Appuhn (2009).
2. Evelyn (1995), p.320.
3. Quoted from Batey (2007), p.5.
4. Batey (2007), p.67.
5. Evelyn (1995), p.320.
6. Ibid., p.384.
7. Ibid., p.324.
8. Ibid., p.298.
9. Ibid., p.327.
10. For a history of the forests under the ancien régime, see Corvol (1987).
11. This and the following quotations are from the French Forest Ordinance of 1669. Available at http://openlibrary.org.

Chapter Seven
1. This and the following quotations are from Carlowitz (2000), chapters I-VII.
2. Seckendorff (1972), p.474.
3. Carlowitz (2000), p.105.
4. Quoted from Totman (1998), p.77.

Chapter Eight
1. For a groundbreaking study on the history of the concept see Worster (1977).
2. Stillingfleet (1775), p.39. Available at http://google.books.com.
3. Ibid., p.40.
4. Quoted from Koerner (1999), p.83.
5. Koerner (1999), p.103.
6. Goethe (1995).
7. This and the following quotations: Herder (1966), p.1.
8. Ibid., p.13.
9. Ibid., p.34.
10. Ibid., p.3.
11. Goethe (1986), p.101.

12. Humboldt (1859), p.79.
13. Haeckel (1866), p.286. Available at http://google.books.com. For a detailed account of Haeckel's book see Richards (2008), Chapter 5.
14. The semantics of 'milieu', 'Umwelt' and 'environment' are profoundly discussed by Spitzer (1948).
15. Goethe (1989), *Italian Journey*, p.23.
16. Carlyle (1899), p.222.

Chapter Nine
1. Hartig (1804).

Chapter Ten
1. This and the following quotations are from Evelyn (1995), pp.126-156.
2. For the history of fossil energy use, see Sieferle (2001).

Chapter Eleven
1. Schlich (1990), p.19.
2. For the whole passage see Pinchot (1998).
3. The story of the Dust Bowl is told by Worster (2004).
4. Leopold (1970), p.262.
5. Duerr (1975), p.36.

Chapter Twelve
1. Meadows (1972), p.158.
2. Ibid., p.29.
3. Basler (1973), pp.74-97.
4. The documents of the 1972 Stockholm conference are available at www.unep.org.
5. The story of the debate at WCC is told by Birch (1993), pp.113-115. The documents of the Bucharest Conference were published in the WCC's journal *Anticipation*, No. 19, November 1974.
6. See Klein (2007).
7. Lovelock (1988), p.30.
8. Ibid., p.199.

Chapter Thirteen
1. See IUCN (1980).
2. Brandt (1980), p.9. The Brandt Report is also available from http://files.globalmarshallplan.org.
3. Ibid., p.14.
4. Ibid., p.9.

5. Ibid., p.17.
6. IFDA (1980), p.10. Available at www.dhf.uu.se
7. WCED (1987), p.356.
8. Brundtland (2002), p.208.
9. WCED (1987), p.43.
10. Ibid., p.1.
11. Earth Charter Commission (2002). Also available at www.earthcharter.org.

Chapter Fourteen
1. Crist (2007), p.40.
2. Ibid., p.34.

Bibliography

Adams, Frank Dawson. 1938. *The Birth and Development of Geological Sciences.* Baltimore: Williams and Wilkins. Available from www.archive.org

Apollo 17 Flight Journal. 1972. Available from www.ehartwell.com

Appuhn, Karl. 2009. *A Forest on the Sea.* Baltimore: John Hopkins University Press.

Basler, Ernst. 1973. *Strategie des Fortschritts. Umweltbelastung, Lebensraumverknappung und Zukunftsforschung.* Frauenfeld, Switzerland: Huber.

Batey, Mavis (ed.). 2007. *A Celebration of John Evelyn.* Sutton, UK: Surrey Gardens Trust.

Birch, Charles. 1993. *Regaining Compassion – for humanity and nature.* Kensington, Australia: New South Wales University Press.

Bosselmann, Klaus. 2008. *The Principle of Sustainability. Transforming Law and Governance.* Farnham, UK: Ashgate.

Brandt, Willy, et al. 1980. *North-South: A Programme for Survival.* London: Pan Books.

Brundtland, Gro Harlem. 2002. *Madam Prime Minister. A Life in Power and Politics.* New York: Farrar, Straus and Giroux.

Carlowitz, Hans Carl von. 2000. *Sylvicultura oeconomica.* Leipzig 1713. Reprint. Freiberg, Germany: TU Bergakademie.

Carlyle, Thomas. 1899. *Goethe (1828).* In: Critical and Miscellaneous Essays, Vol. I. London: Chapman and Hall.

Carson, Rachel. 2000. *Silent Spring* (first published 1962). London: Penguin Classics.

Corvol, Andrée. 1987. *L'Homme aux Bois.* Paris: Editions Fayard.

Crist, Eileen. 2007. *Beyond the Climate Crisis: A Critique of Climate Change Discourse.* In: *Telos* 141.

Cusa, Nicholas of. 1981. *On Learned Ignorance.* A Translation and an Appraisal of *De Docta Ignorantia* by Jasper Hopkins. Minneapolis: Arthur Banning Press. Avaialable from http://my.pclink.com

Descartes, René. 1998. *Discourse on Method and Meditations on First Philosophy.* Trans. Donald A. Cress. Indianapolis: Hackett. Available from www.naderlibrary.com

Duerr, William A. and Rumsey, Fay (eds.). 1975. *Social Sciences in Forestry: A Book of Readings.* Philadelphia: W.B. Saunders.

Earth Charter Commission. 2002. *Earth Charter: Values and Principles for a Sustainable Future*. Brochure.

Evelyn, John. 1995. *Sylva, or a Discourse of Forest-Trees and the Propagation of Timber*. In: The Writings of John Evelyn. Ed. by Guy de la Bédoyère. Woodbridge, UK: Boydell Press. The text of *Sylva* is available from www.gutenberg.org

Evelyn, John. 1995. *Fumifugium or The Inconvenience of the Aer and Smoak of London*. In: The Writings of John Evelyn. Ed. by Guy de la Bédoyère. Woodbridge, UK: Boydell Press.

French Forest Ordinance of 1669. 1883. In: Brown, John Croumbie. *The French Forest Ordinance of 1669*. Edinburgh: Oliver and Boyd.

Freud, Sigmund. 1961. *Civilization and Its Discontents*. Trans. James Strachey. New York: W.W. Norton.

Goethe, Johann Wolfgang von. 1986. *Essays on Art and Literature*. Ed. by John Gearey. Trans. Ellen von Nordroff. Goethe's Collected Works, Vol. 3. New York: Suhrkamp Publishers.

Goethe, Johann Wolfgang von. 1989. *Wilhelm Meister's Apprenticeship*. Ed. and trans. by Eric A. Blackall. Goethe's Collected Works, Vol. 9. New York: Suhrkamp.

Goethe, Johann Wolfgang von. 1989. *Italian Journey*. Trans. Robert R. Heitner. Goethe's Collected Works, Vol. 6. New York: Suhrkamp.

Goethe, Johann Wolfgang von. 1995. *Scientific Studies*. Ed. and trans. by Douglas Miller. The Collected Works Vol. 12. Princeton: Princeton Paperbacks.

Haeckel, Ernst. 1866. *Generelle Morphologie der Organismen*. Berlin: Georg Reimer.

Hartig, Georg Ludwig. 1804. *Anweisung zur Taxation der Forste*. 2. Auflage. Giessen.

Herder, Johann Gottfried von. 1966. *Outlines of a Philosophy of the History of Man*. Trans. T. Churchill (first published in London, 1800). New York: Bergman Publishers

Humboldt, Alexander von. 1859. *Cosmos: A Sketch of a Physical Description of the Universe*. Vol. 1, trans. E. C. Otté. Available from www.books.google.com.

IFDA (International Foundation for Development Alternatives). 1980. *Building Blocks for Alternative Development. A Progress Report*. Zug, Switzerland: IFDA.

IUCN (International Union for Conservation of Nature and Natural Resources). 1980. *World Conservation Strategy. Living Resource Conservation for Sustainable Development*. Gland, Switzerland: IUCN.

Kelley, Kevin W. 1988. *The Home Planet*. Reading, MA: Addison-Wesley.

Kepler, Johannes. 1967. *Kepler's Somnium. The Dream, or Posthumous Work on Lunar Astronomy*. Trans. Edward Rosen. Madison WI: University of Wisconsin Press.

Kittler, Friedrich A. 2001. 'Computer Graphics'. Trans. Sara Ogger, in *Grey Room*, No. 2.

Klein, Naomi. 2007. *The Shock Doctrine. The Rise of Disaster Capitalism*. New York: Metropolitan Books.

Koerner, Lisbeth. 1999. *Linnaeus: Nature and Nation*. Cambridge, MA: Harvard University Press.

Leclerc, Eloi. 1977. *The Canticle of Creatures*. Trans. Matthew J. O' Connell. Chicago: Franciscan Herald Press.

Leopold, Aldo. 1970: *A Sand County Almanac*. With Essays on Conservation from Round River. New York: Oxford University Press.

Lovelock, James. 1988. *The Ages of Gaia. A Biography of Our Living Earth*. New York/London: W.W. Norton.

Maslow, Abraham H. 1987. *Motivation and Personality*. Reading, MA: Addison-Wesley.

Meadows, Donella H., Meadows, Dennis L., Randers, Jorgen et al. 1972. *The Limits to Growth. A Report to the Club of Rome's Project on the Predicament of Mankind*. London: Earth Island.

Merchant, Carolyn. 1980. *The Death of Nature: Women, Ecology and the Scientific Revolution*. San Francisco: Harper & Row.

Niavis, Paulus. 1953. *Iudicium Jovis oder das Gericht der Götter über den Bergbau*. Ed. by Paul Krenkel. Berlin: Freiberger Forschungshefte.

Pinchot, Gifford. 1998. *Breaking New Ground*. Washington D.C.: Island Press.

Poole, Robert. 2008. *Earthrise. How Man First Saw the Earth*. New Haven / London: Yale University Press.

Pye-Smith, Charlie. 1994. *The Wealth of Communities*. With a foreword by HRH The Prince of Wales. London: Earthscan Publications.

Richards, Robert J. 2008. *The Tragic Sense of Life. Ernst Haeckel and the Struggle over Evolutionary Thought*. Chicago: The University of Chicago Press.

Rifkin, Jeremy. 2004. *The European Dream*. New York: Jeremy P. Tarcher/Penguin.

Schlich, William. 1990. *Schlich's Manual of Forestry*. Vol I, 3rd ed. , Oxford 1922. Reprint: Delhi: Periodical Experts Book Agency.

Seckendorff, Veit Ludwig von. 1972. *Teutsche Fürstenstaat* (1656). Aalen: Scientia.

Sieferle, Rolf Peter. 2001. *The Subterranean Forest*. Cambridge, UK: White Horse Press.

Spinoza, Benedict de. 1883. *The Ethics*. Trans. R.H.M. Elwes.

Spitzer, Leo. 1948. 'Milieu and Ambiance' in *Essays in Historical Semantics*. New York: S.F. Vanni. pp.179-316.

Stillingfleet, Benjamin. 1775. *Miscellaneous Tracts Relating to Natural History, Husbandry and Physick*. London. (English translation of treatises by Linnaeus and his students).

Totman, Conrad D. 1998. *The Green Archipelago. Forestry in Preindustrial Japan.* Athens, OH: Ohio University Press.

WCED (World Commission on Environment and Development). 1987. *Our Common Future. (Brundtland Report).* Oxford / New York: Oxford University Press.

Weart, Spencer R. 2008. *The Discovery of Global Warming.* Cambridge, MA: Harvard University Press.

Whitney, William. 1905. *Atharva-Veda-Samitha*, Vol. II. Delhi.

Worster, Donald. 1977. *Nature's Economy. A History of Ecological Ideas.* San Francisco: Sierra Club Books.

Worster, Donald. 2004. *Dust Bowl. The Southern Plains in the 1930s.* (25th Anniversary Edition). New York: Oxford University Press.

Index

We publish a wide range of books on ecological
and cultural issues, including gardening, eco-building, economics,
politics and green living. For a complete list, please visit our website:

www.greenbooks.co.uk